Recovery and Utilization of Metallurgical Solid Waste

Edited by Yingyi Zhang

Published in London, United Kingdom

IntechOpen

Supporting open minds since 2005

Recovery and Utilization of Metallurgical Solid Waste
http://dx.doi.org/10.5772/intechopen.76826
Edited by Yingyi Zhang

Contributors
Hu Long, Dong Liu, Lie-Jun Li, Ming-Hua Bai, Yanzhong Jia, Wensheng Qiu, Andrey Dmitriev, Július Strigáč, Nadežda Števulová, Jozef Mikušinec, Ľudovít Varečka, Daniela Hudecová, Elena Brandaleze, Angélica Deus, Rosemary Marques de Almeida Bertani, Leonardo Büll, Dirceu Fernandes, Guilherme Meirelles, Anelisa Soares, Lais Moreira, Yingyi Zhang

First published in London, United Kingdom, 2019 by IntechOpen
IntechOpen is the global imprint of INTECHOPEN LIMITED, registered in England and Wales, registration number: 11086078, The Shard, 25th floor, 32 London Bridge Street
London, SE19SG - United Kingdom
Printed in Croatia

British Library Cataloguing-in-Publication Data
A catalogue record for this book is available from the British Library

Additional hard copies can be obtained from orders@intechopen.com

Recovery and Utilization of Metallurgical Solid Waste
Edited by Yingyi Zhang
p. cm.
Print ISBN 978-1-78985-101-4
Online ISBN 978-1-78985-102-1

Meet the editor

Yingyi Zhang (1985–) is an associate professor at Anhui University of Technology. He has presided and participated in more than 20 research projects. He is mainly engaged in research of the development of new ironmaking technology, comprehensive utilization of metallurgical solid waste, and the preparation of high-temperature coatings. With more than 50 publications and 30 patents, Prof. Zhang has dedicated himself to the comprehensive utilization of metallurgical solid waste for some time. In recent years, he has participated in two national science and technology support programs, one international science and technology cooperation and exchange program, and six national natural science foundation programs. In addition, he has hosted one Chinese postdoctoral science foundation program, one national natural science foundation program, and two national key research and development programs.

Contents

Preface

The coordinated development of resources, environment, energy and population is a major social problem in today's world. A lot of metallurgical solid wastes have not been timely and effectively recycled, resulting in serious problems of environmental pollution and waste of resources. As a result, large scale comprehensive utilization technologies have been initiated, including slag dry granulation technology, steel slag cement technology, slag wool technology and slag waste heat recovery technology, etc. To help support the effort, various governmental and non-governmental agencies established funding for metallurgical solid waste comprehensive utilization technology. The comprehensive utilization of metallurgical solid waste has attracted worldwide attention. It is an effective way to improve the utilization efficiency of resources and the added value of products by using the scientific metallurgical solid waste recycling methods.

This book is divided into five chapters. The first chapter is written by Prof. Striga Julius, et al. This chapter focuses on the antimicrobial efficiency of blast furnace slags for applications in building materials and products. The second chapter is written by Dr. Brandaleze Elena et al. In this chapter, the treatment and recycling techniques of BOF and EAF slags are described. The BOF slags are applied in buildup, foaming or slag splashing to prolong refractory lining life. Also EAF slags are commonly used to avoid refractory wear and reduce energy consumption. The BOF and EAF slag are also used in cement production and building materials. The third chapter is written by Dr. Deus Angélica et al. This chapter aims to emphasize the application of slags in soil, and the effect of slags on nutrient content, such as phosphorus, calcium and magnesium, some micronutrients and silicon. In addition, the utilization of these residues in tropical soils is also discussed. The fourth chapter is written by Dr. Long Hu et al. In this chapter, the physical-chemical properties and mineralogical phases of converter sludge are characterized, and different recycling technologies are introduced. The metalized pellet producing process is mentioned, whereby green pellets made from iron-bearing sludge are dried and preheated firstly, and then reduced at high temperature in a rotary kiln or a rotary hearth furnace (RHF) to obtain direct reduced iron (DRI). The fifth chapter is written by Prof. Dmitriev Andrey et al. In this chapter, the red mud of the bauxite process is used to produce iron-ore sinter and pellets, the effect of red mud contents on metallurgical properties of sinter and pellets is investigated, and the main technical and economic indicators of blast furnace smelting are also analyzed.

This book intends to provide the reader with a comprehensive overview of the metallurgical solid wastes comprehensive utilization technology. The comprehensive utilization methods of metallurgical solid wastes are emphatically described, such as blast furnace slag, steel slag, converter sludge and red mud.

During the writing process, the book was helped in many ways. Thanks to Prof. Striga Julius, Prof. Dmitriev Andrey, Dr. Brandaleze Elena, Dr. Deus Angelica,

Dr. Long Hu and others for their hard work and support. Thanks also to Ms. Dolores Kuzelj, Author Service Manager, for her help and guidance, and thanks to colleagues at Anhui University of Technology for their strong support.

Yingyi Zhang
Anhui University of Technology,
Anhui Sheng, China

Section 1

Review of Comprehensive Utilization of Metallurgical Solid Waste

Chapter 1

Introductory Chapter: Metallurgical Solid Waste

Yingyi Zhang

1. Utilization status of metallurgical solid waste resource

The sustainable development of resources and energy is an inevitable trend of social development. With the rapid development of metallurgical industry, a large amount of metallurgical solid waste is produced. However, a lot of metallurgical solid wastes have not been timely and effectively recycled, resulting in serious problems of environmental pollution and resource wastage, such as heavy metal pollution in air, water, soil, and plant system. According to the characteristics of metallurgical industry, this book introduces the main types, sources, and characteristics of metallurgical solid waste. The application and treatment methods of blast furnace slag, converter slag, and electric furnace slag in building materials and ceramics industry are mainly introduced. The comprehensive utilization technology of Bayer process-produced red mud and converter sludge was also investigated.

Blast furnace slag (BFS) is a by-product of iron-making, which is formed by the combination of iron ore with limestone flux. When the molten slag is rapidly cooled by water, a large amount of granulated and amorphous blast furnace slag is produced, and the physical and chemical properties of these slags mainly depend on the production process. It is worth noting that a lot of water and heat are wasted in this process. The main chemical components of these slags are silica (SiO_2), alumina (Al_2O_3), and lime (CaO), which are the main components of cement and CaO-Al_2O_3-SiO_2 (CAS) glass ceramics [1, 2]. Therefore, the granulated blast furnace slag (GBFS) is usually used as feedstock for cement and glass ceramics manufacturing. The CaO content of blast furnace slag is about 35–56% [3], which can hydrate with water to form cementitious pozzolanic reaction products. When these slags are ground to a finer size, they can be utilized in the production of Portland slag cement [4]. The typical TEM micrograph of hydrated Portland cement (a) and photograph of the glass-ceramics (b) just fabricated from blast furnace slag are shown in **Figure 1**.

Steel slags are by-products of steel production, which are produced during the electric arc furnace process in steel-making, converter steel-making, and the secondary refining of steel, respectively [7]. In the process of steel-making, steel slag production accounts for about 15% of steel output. The steel slag annual production worldwide is about 130 million tons which are mainly electrical arc furnace slag (EAF), basic oxygen furnace (BOF) slags, and ladle furnace basic slag (LFS) [8]. The main chemical components of steel slag include silica (SiO_2), alumina (Al_2O_3), lime (CaO), magnesia (MgO), ferrous oxide (FeO), and hematite (Fe_2O_3), as shown in **Table 1**. The mineral components of steel slag mainly consist of olivine, hydraulic calcium silicate (β-Ca_2SiO_4 (C_2S), Ca_3SiO_5 (C_3S)), non-hydraulic calcium silicate (γ-Ca_2SiO_4 (C_2S), $CaSiO_3$ (CS)), tetra-calcium aluminoferrite (C_4AF), dicalcium ferrite (C_2F), and free CaO/MgO [15, 16]. As we all know, C_2S and C_3S are the main

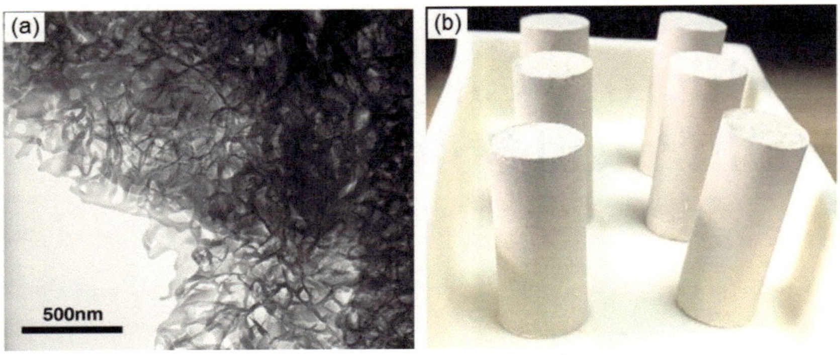

Figure 1.
TEM micrograph of hydrated Portland cement (a) and photograph of the glass-ceramics (b) just fabricated from blast furnace slag [5, 6].

Reference	Type	CaO	SiO_2	Al_2O_3	MgO	Fe_2O_3
[9]	BOFS	39.08	12.47	6.87	10.57	19.48
[10]		61.21	24.92	1.83	4.89	3.04
[11]		47.7	13.3	3.0	6.4	24.4
[12]	EAFS	38.8	14.1	6.7	3.9	20.3
[13]		24.4	15.4	12.2	2.9	—
[9]	LFS	57.55	6.21	23.17	5.04	3.55
[14]		50.5–57.5	12.6–19.8	4.3–18.6	7.5–11.9	1.6–3.3
[12]		42.5	31.9	22.9	12.6	1.1

Table 1.
The main chemical compositions of steel slag (wt%).

components of cement. Therefore, the steel-making slag also has certain cementitious properties and is used as feedstock for cement manufacturing, which is widely used in the construction industry. **Figure 2** shows the application of electric furnace slag concrete in construction.

Red mud (bauxite residue) is an alkaline solid waste residue generated from the Bayer process [17]. Presently, when producing a ton of alumina via Bayer process, about 0.8–2.0 t of red mud residues are produced, which mainly depends on the properties of raw material and production process conditions [18]. The global annual production of high-alkalinity red mud is about 120 million tons, and the global accumulation of red mud is about 2.7 billion tons [19]. The typical chemical compositions and mineralogical components of red mud are shown in **Table 2**. It can be seen that the red mud has a high pH or alkalinity, the aluminum compounds mainly consist of gibbsite ($Al(OH)_3$), aluminous goethite (α-(Fe,Al)OOH), boehmite (AlOOH) or diaspore (AlOOH), and the iron oxide mainly consists of hematite (Fe_2O_3) and aluminous goethite (α-(Fe,Al)OOH). However, large quantities of red mud are discarded as waste, and have not been effectively developed and utilized. This leads to serious soil, air, and water pollution and takes up a lot of space [20]. At present, most of the red mud is directly placed in landfill, deep sea, and storage in settling ponds, as shown in **Figure 3**. Despite the harmful impact that these methods pose on our environment, the risks of failure of a poorly engineered storage dam can result in even greater social and economic damage. In addition, the

Figure 2.
The application of EAF slag concrete in pavements [7].

Chemical name	Composition/ wt%	Mineralogical name	Chemical formula	Composition/ wt%
Fe_2O_3	5–60%	Sodalite	$3Na_2O{\cdot}3Al_2O_3{\cdot}6SiO_2{\cdot}Na_2SO_4$	4–40%
Al_2O_3	5–30%	Aluminous goethite	α-(Fe,Al)OOH	10–30%
TiO_2	0.3–15%	Hematite (iron oxide)	Fe_2O_3	10–30%
CaO	2–20%	Silica	SiO_2	5–20%
SiO_2	3–50%	Tricalcium aluminate	$3CaO{\cdot}Al_2O_3{\cdot}6H_2O$	2–20%
Na_2O	1–10%	Boehmite	AlO(OH)	0–20%
		Titanium dioxide	TiO_2	2–15%
		Muscovite	$K_2O{\cdot}3Al_2O_3{\cdot}6SiO_2{\cdot}2H_2O$	0–15%
		Calcium carbonate	$CaCO_3$	2–10%
		Gibbsite	$Al(OH)_3$	0–5%
		Kaolinite	$Al_2O_3{\cdot}2SiO_2{\cdot}2H_2O$	0–5%

Table 2.
Chemical composition and mineralogical components of red mud (wt%).

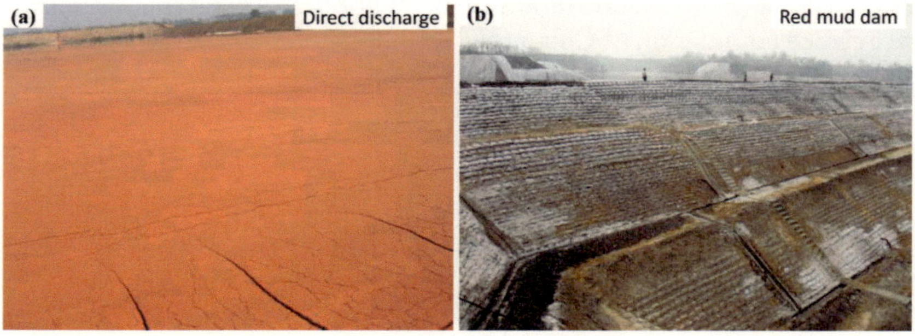

Figure 3.
Typical views of the red mud (a) and red mud dam (b).

red mud has a high concentration of aluminum compound, sodium aluminate, and iron oxide sodium, which limits the application of red mud in cement and ceramic industry (**Tables 2** and **3**). Therefore, the comprehensive utilization of red mud residue is an urgent problem in alumina industry.

Reference	Fe_2O_3	Al_2O_3	CaO	SiO_2	TiO_2	Na_2O	MgO	Cr_2O_3	V_2O_5	LOI
[21]	43.59	18.45	11.38	6.0	5.54	1.82	0.35	0.27	0.17	11.7
[22]	28.30	17.67	20.88	8.34	7.34	2.29	0.65	—	—	13.88
[17]	54.8	14.8	2.5	6.4	3.7	4.8	—	—	0.38	9.5
[23]	32.52	18.42	16.74	8.34	6.75	3.59	—	—	—	13.64

Table 3.
The chemical composition of red mud (wt%).

2. Conclusions

The sustainable development of resources and energy is an inevitable trend of social development. The comprehensive utilization of metallurgical solid waste is still a worldwide problem. Because of the limitations of current technology and consumption levels, a large amount of metallurgical solid waste has not been exploited effectively. The granulated blast furnace slag and steel slags are usually used as feedstock for cement and glass ceramics manufacturing. However, the traditional water quenching slag process wastes a lot heat of slag, polluted environment, and consumed water resources. Therefore, it is very important to develop technology for the utilization of waste heat from blast furnace slag and steel slag. Industrial storage is not the only way to solve the problem of comprehensive utilization and pollution of red mud. The recovery of valuable metals from red mud faces many technical problems, which seriously hinder the development of the metallurgical industry. In addition, applying red mud as construction material like cement or soil ameliorant faces the problem of Na, Cr, and As leaching into the environment. So, we must reduce the recycling process costs and energy consumption, promote the recovery of valuable metals, optimize complex processes, and develop new processes.

Acknowledgements

This work was supported by the National Key Research and Development Program of China (2017YFB0603800 & 2017YFB0603802) and the National Natural Science Foundation of China (No.51604049).

Author details

Yingyi Zhang
School of Metallurgical Engineering, Anhui University of Technology, Maanshan, Anhui Province, P. R. China

*Address all correspondence to: zhangyingyi@cqu.edu.cn

References

[1] Itoh T. Rapid discrimination of the character of the water-cooled blast furnace slag used for Portland slag cement. Journal of Materials Science. 2004;**39**(6):2191-2193. DOI: 10.1023/B:JMSC.0000017785.44922.30

[2] Zhang WT, He F, Xie JL, Liu XQ, Fang D, Yang H, et al. Crystallization mechanism and properties of glass ceramics from modified molten blast furnace slag. Journal of Non-Crystalline Solids. 2018;**502**:164-171. DOI: 10.1016/j.jnoncrysol.2018.08.024

[3] Kumar S, Kumar R, Bandopadhyay A, Alex TC, Kumar BR, Das SK, et al. Mechanical activation of granulated blast furnace slag and its effect on the properties and structure of Portland slag cement. Cement & Concrete Composites. 2008;**30**(8):679-685. DOI: 10.1016/j.cemconcomp.2008.05.005

[4] Sekhar DC, Nayak S. Utilization of granulated blast furnace slag and cement in the manufacture of compressed stabilized earth blocks. Construction and Building Materials. 2018;**166**:531-536. DOI: 10.1016/j.conbuildmat.2018.01.125

[5] Taylor R, Richardson IG, Brydson RMD. Composition and microstructure of 20-year-old ordinary Portland cement-ground granulated blast-furnace slag blends containing 0 to 100% slag. Cement and Concrete Research. 2010;**40**(7):971-983. DOI: 10.1016/j.cemconres.2010.02.012

[6] Gao HT, Liu XH, Chen JQ, Qi JL, Wang YB, Ai ZR. Preparation of glass-ceramics with low density and high strength using blast furnace slag, glass fiber and water glass. Ceramics International. 2017;**44**(6):6044-6053. DOI: 10.1016/j.ceramint.2017.12.228

[7] Fuente-Alonso JA, Ortega-López V, Skaf M, Aragón Á, San-José JT. Performance of fiber-reinforced EAF slag concrete for use in pavements. Construction and Building Materials. 2017;**149**:629-638. DOI: 10.1016/j.conbuildmat.2017.05.174

[8] Zomeren AV, VanderLaan SR, Kobesen HBA, Huijgen WJJ, Comans RNJ. Changes in mineralogical and leaching properties of converter steel slag resulting from accelerated carbonation at low CO_2 pressure. Waste Management. 2011;**31**(11):2236-2244. DOI: 10.1016/j.wasman.2011.05.022

[9] Mahoutian M, Chaallal O, Shao Y. Pilot production of steel slag masonry blocks. Canadian Journal of Civil Engineering. 2018;**45**(7):537-546. DOI: 10.1139/cjce-2017-0603

[10] Ghouleh Z, Guthrie RIL, Shao Y. High-strength KOBM steel slag binder activated by carbonation. Construction and Building Materials. 2015;**99**:175-183. DOI: 10.1016/j.conbuildmat.2015.09.028

[11] Waligora J, Bulteel D, Degrugilliers P, Damidot D, Potdevin JL, Measson M. Chemical and mineralogical characterizations of LD converter steel slags: A multi-analytical techniques approach. Materials Characterization. 2010;**61**(1):39-48. DOI: 10.1016/j.matchar.2009.10.004

[12] Tossavainen M, Engstrom F, Yang Q, Menad N, Lidstrom Larsson M, Bjorkman B. Characteristics of steel slag under different cooling conditions. Waste Management. 2007;**27**(10):B1335-B1344. DOI: 10.1016/j.wasman.2006.08.002

[13] Luxan MP, Sotolongo R, Dorrego F, Herrero E. Characteristics of the slags produced in the fusion of scrap steel by electric arc furnace. Cement and Concrete Research. 2000;**30**(4):517-519. DOI: 10.1016/S0008-8846(99)00253-7

[14] Setién J, Hernández D, González JJ. Characterization of ladle furnace basic slag for use as a construction material. Construction and Building Materials. 2009;**23**(5):1788-1794. DOI: doi.org/10.1016/j.conbuildmat.2008.10.003

[15] Maslehuddin M, Sharif AM, Shameem M, Ibrahim M, Barry MS. Comparison of properties of steel slag and crushed limestone aggregate concretes. Construction and Building Materials. 2003;**17**(2):105-112. DOI: 10.1016/S0950-0618(02)00095-8

[16] Tsakiridis PE, Papadimitriou GD, Tsivilis S, Koroneos C. Utilization of steel slag for Portland cement clinker production. Journal of Hazardous Materials. 2008;**152**(2):805-811. DOI: 10.1016/j.jhazmat.2007.07.093

[17] Samal S, Ray AK, Bandopadhyay A. Proposal for resources, utilization and processes of red mud in India-a review. International Journal of Mineral Processing. 2013;**118**(30):43-55. DOI: 10.1016/j.minpro.2012.11.001

[18] Liu WC, Yang JK, Xiao B. Review on treatment and utilization of bauxite residues in China. International Journal of Mineral Processing. 2009;**93**(3-4):220-231. DOI: 10.1016/j.minpro.2009.08.005

[19] Klauber C, Gräfe M, Power G. Bauxite residue issues: II options for residue utilization. Hydrometallurgy. 2011;**108**(1-2):11-32. DOI: 10.1016/j.hydromet.2011.02.007

[20] Li RB, Zhang TA, Liu Y, Lv GZ, Xie LQ. Calcification-carbonation method for red mud processing. Journal of Hazardous Materials. 2016;**316**:94-101. DOI: 10.1016/j.jhazmat.2016.04.072

[21] Samouhos M, Taxiarchou M, Pilatos G, Tsakiridis PE, Devlin E, Pissas M. Controlled reduction of red mud by H_2 followed by magnetic separation. Minerals Engineering. 2017;**105**(1):36-43. DOI: 10.1016/j.mineng.2017.01.004

[22] Liu DY, Wu CS. Stockpiling and comprehensive utilization of red mud research progress. Materials. 2012;**5**(7):1232-1246. DOI: 10.3390/ma5071232

[23] Li XB, Wi X, Liu W, Liu GH, Peng ZH, Zhou QS, et al. Recovery of alumina and ferric oxide from Bayer red mud rich in iron by reduction sintering. Transactions of Nonferrous Metals Society of China. 2009;**19**(5):1342-1347. DOI: 10.1016/S1003-6326(08)60447-1

Section 2

Comprehensive Utilization of Blast Furnace Slag

Chapter 2

Antimicrobial Efficiency of Metallurgical Slags Suitable for Construction Applications

Július Strigáč, Nadežda Števulová, Jozef Mikušinec, Ľudovít Varečka and Daniela Hudecová

Abstract

The chapter deals with studying antimicrobial efficiency of granulated blast-furnace slag with fineness of 340 (1Sa) and 520 m^2/kg (1Sb), air-cooled blast-furnace slag (2S), demetallized steel slag (3S), calcareous ladle slag (4S) and copper slag (5S), respectively. The efficiency has been tested on G^+ bacteria—*Staphylococcus aureus*, *Bacillus subtilis*, *Micrococcus luteus*; G^- bacteria—*Pseudomonas aeruginosa*, *Escherichia coli*, *Serratia marcescens*; yeasts—*Rhodotorula glutinis*, *Candida albicans*; filamentous fungi—*Penicillium funiculosum*, *Aspergillus niger*, *Alternaria alternata*, *Chaetomium globosum*, *Cladosporium herbarum*, *Trichoderma viride*. The efficiency has been determined by dilution methods in agar media for that reason the resulting concentration of slags has been 10, 20, 40 and 60%, respectively. The antibacterial efficiency decreased as follows: S4 > S3 > S2 > S1a = S1b > S5, whereas anti-yeast efficiency decreased as follows: S4 > S1a = S1b = S3 > S2 > S5. Filamentous fungi were selectively sensitive to slags, that way there is approximate order of efficiency S4 > S3 = S1a = S1b > S5 > S2. Application of metallurgical slags into construction materials provides them increasing biodegradation resistance.

Keywords: metallurgical slags, antimicrobial efficiency, bacteria, yeasts, filamentous fungi

1. Introduction

Biocorrosion of construction materials is a significant problem wherever the conditions suitable for microorganisms occur. Biocorrosion is caused by varied biogenic acids, as well as by H_2S and NH_3, which result from metabolic activities of microorganisms [1–4]. These corrosive metabolites react with calcareous components of construction materials, finally leading to the biodeterioration.

The effects of different compositions of concrete and added nutrients on fungal colonization have been studied in [5]. Fungal strains belonging to *Cladosporium*, *Alternaria*, *Epicoccum*, *Mucor*, *Fusarium*, *Penicillium*, *Trichoderma*, and *Pestalotiopsis* were isolated from fouled concrete structures and used to inoculate mortar tiles varying in cement composition, water-to-cement ratio, supplementary cementitious material additions, and surface roughness. The strong positive relationship has been observed between the tile water-to-cement ratio and the amount of biofouling.

Blast-furnace slag has been added in different amounts of 10, 25, and 50%, and the results showed, that it possessed only a small antifungal effect [5].

To the contrary, it has been determined [6] that granulated blast-furnace slag (GBFS), as well as blast-furnace cements with GBFS ≥65 wt.% have fungistatic properties. Therefore, the utilization of fungistatic GBFS, blast-furnace cements CEM III/A 32,5 N (with GBFS content of 65 wt.%), CEM III/B 32,5 N, and CEM III/C 32,5 N provides long-term fungistatic protection [6].

Effect of pig iron slag (IrS) particles on soil physicochemical, biological and enzyme activities, was studied in [7]. Contamination of IrS particles altered these soil properties. While the pH value of soil increased slightly, the electrical conductivity, phosphorus, potassium and carbon contents increased substantially in polluted soil. Soil contamination with pig IrS caused substantial decrease in microbial population as well as cease of enzyme activities such as dehydrogenase and protease. Soil is a dynamic system in which continuous interaction takes place between soil minerals, organic matter and organisms. Each of these three major soil components influences the physicochemical and biological properties of terrestrial system. Since soil enzyme activities are very sensitive to pollution, enzymes have been suggested as potential indicators or monitoring tools to assess soil quality and bioremediation activities. The aim in the study [7] was to determine the effect of pig IrS on soil physicochemical, biological and enzymatic activities of dehydrogenase, protease, cellulase and amylase. Soil microbial populations such as including bacterial and fungal populations were enumerated by serial dilution technique. The observations revealed that contamination with IrS particles led to increase in pH of the test soil. This indicates the alkaline nature of the disposed waste contaminants. Reduced bacterial and fungal population in polluted soil may be due to the toxic effects of IrS particles on microbial population. This may be because of oxidative stress caused by iron sledge particles or their interference in osmotic balance [7]. Pollution of the soil environment with heavy metals also negatively influenced the soil microbial properties such as basal soil respiration rate and enzyme activities depending on the soil pH, organic matter content and other chemical properties. Enzymes are strongly connected with important soil characteristics such as organic matter, physical properties, microbial activity or biomass. They are the sensitive indicators of soil quality. The study clearly indicated that dumping of pig IrS on soil, altered its physicochemical and biological properties and inhibited enzymatic activities such as dehydrogenase and protease. Very significant inhibition of dehydrogenase and protease activities in polluted soil indicates the alterations in oxidation-reduction activities of enzymes released from microorganisms [7].

The use of steelmaking slag for sewage sludge stabilization was studied in [8]. The objective was the examination of the stabilization of sewage sludge by the addition of a steel slag as an alkaline material and the investigation of the effect of stabilization time on the properties of the produced mixtures. The alkaline material used was a by-product from steelmaking refining processes in ladle furnaces, the steelmaking slag that presented high calcium content. The mixtures of sewage sludge and slag in various ratios (2.5–20%) were prepared and stabilized for 48 days. The determination of pathogens removal rate, pH, moisture content and mineralogical phases present was carried out for monitoring of the mixtures properties. The slag addition resulted in increasing mixture pH (exceeding 12), in the increase of total solids content and in the decrease of volatile percentage, which was more pronounced at the highest alkaline slag dosage. In addition, effective pathogens removal was observed in the mixtures containing more than 10% slag, due to the high pH values of mixtures. The addition of slag at lower dosages was not as effective as the highest dosage. Sewage sludge has been utilized for agriculture and horticulture for several years and represents a good source of nutrients for plant growth and soil conditioner to improve soil physical properties. One of the alternative methods for sludge hygienisation is chemical stabilization with lime at a pH value equal or above 12 for 3 months. Lime maintains high pH values of mixtures, removing microbial communities in sludge. An alternative alkaline agent for sludge stabilization, other than lime, is the residue produced during mechanical treatment of steel in steelmaking processes (steelmaking slag). Steelmaking slag is a by-product from steel refining processes in ladle furnaces and contains a high calcium amount, representing a medium with strong alkaline properties. The steelmaking slag used represents a high alkaline material with the following average composition, on a dry basis: CaO 53 wt.%; SiO_2 18 wt.%; MgO 4.5 wt.%; Al_2O_3 3 wt.%; MnO 2 wt.%; iron oxides 7.5 wt.% [8]. The alkaline agent addition in sewage sludge is supposed to contribute to sewage sludge stabilization through the pH value increase toward highly alkaline values (>11), which results in the destruction of pathogens. Furthermore, calcium cations may react with sulfur containing substances in the sludge, resulting simultaneously to odor reduction. The raw sewage sludge presented high moisture content, reaching up to 83% and pH values in the neutral range. The pH values of alkaline sewage sludge mixtures were higher than the pH values of raw sludge sample, throughout the whole period of 48 days. The highest pH values (exceeding 12) were measured for alkaline sewages with the highest content of alkaline medium, i.e., containing 10 and 20% slag. Lower pH values, from about 7.5 up to 11.0, were measured for the samples with the lower content of the stabilization alkaline material. In general, the higher the slag content, the higher the pathogens removal rate and the lower the required time for stabilization. The sample with slag addition of 20% presented negligible microbial content from the early stabilization days. The lower stabilization efficiencies were measured in the sludge mixtures containing slag contents lower than 5%; however, extended stabilization times resulted in the efficient removal of pathogens even for the samples with a low slag content [8]. In conclusion, the low-cost steelmaking slag, can be used as the efficient alkaline medium alternative to lime, for the sewage sludge stabilization. High slag additions, up to 20%, were able to offer a mixture with a high pH, and a low content of moisture, volatiles and pathogens. The produced mixtures could have several applications, i.e., soil amendment, daily cover in sanitary landfills, restoration of abandoned mines, etc. Comparison of slag to lime and limestone revealed that slag may be as efficient as the most conventional alkaline mediums for sludge stabilization; however, the benefits of the latter over lime, such as integrated use of a solid waste, availability and low cost, indicate that steelmaking slag could become an efficient medium for sludge stabilization [8].

The effect of metallurgical slag on microbiological activity in Pseudogley was studied in [9]. The aim was to investigate the effect of Ca-containing metallurgical slag from iron-producing factory, comparing to the other lime materials and fertilizers on microbiological activity in acid pseudogley type of soil. Metallurgical slag and certain melioration measures did not show significant effect on number of microorganisms in Pseudogley, although the activity of dehydrogenase was significantly high in combined treatments with slag. Based on the obtained results [9], the studied metallurgical slag can be utilized for increasing microbiological activity and fertility of acid soils. The main task of every agricultural production is increasing and maintaining soil fertility. For the majority of pseudogley soils, characterized by high soil acidity, it is necessary to apply Ca-containing materials—calcifiers. Agricultural liming materials increase soil pH and thereby affect the activity and composition of microbial populations. In acid soils, liming can create better environmental conditions for the development of acid-intolerant microorganisms resulting in increased microbial activity. Along with other lime materials (ground limestone, saturated lime, etc.), metallurgical slag can be of great importance [9]. The presence of slag influenced on the decrease of microorganisms number in Pseudogley. Activity of dehydrogenase (DHA) is higher in neutral soils. Higher values of DHA from the studied soil comparing to its control indicate that application of melioration measures and the slag increased its activity. Treatments with slag showed significant activity of DHA in the soil (three times higher comparing to control). Higher activity of DHA indicates higher intensity of respiration, thus, higher intensity of mineralization of soil fresh organic matter and humus. Liming increases soil pH and has a positive and significant effect on microorganisms growth and DHA activity in soil, especially in spring. Thus, the changes in soil pH significantly affect the rate of soil C and N cycling and soil productivity. Based on the obtained results, the studied metallurgical slag of the standardized chemical properties can be utilized for increasing microbiological activity and fertility of acid soils [9].

The mortar composition contains cement, a latent hydraulic material, a pozzolan material and aggregate and further contains non-iron metal refining slag aggregate as a part of the whole of the aggregate to provide a mortar having both excellent acid resistance and anti-microbial action suppressing the activity of sulfur oxidizing bacteria and also excellent in adhesion and crack resistance [10]. The latent hydraulic material is GBFS. The pozzolan material contains one or more kinds of fly ash, pulp sludge incineration ash, silica fume, waste glass powder and sewerage sludge incineration ash. The non-iron refining slag aggregate is one or more kinds of zinc, copper and lead slag.

Metallurgical slags can be advantageously used as an agricultural liming material, as a favorable source of minor nutrients. Fertilizers are primarily valued for their ability to supply nutrients. Plants use these nutrients to make components for plant growth. The main chemicals must be supplied to plants that are called primary nutrients are nitrogen, phosphorus, and potassium [11]. A fertilizer containing all three nutrients is a balanced fertilizer. Plants also require the secondary nutrients, calcium, magnesium, and sulfur, plus very small amounts of the micronutrients boron, copper, chlorine, iron, manganese, molybdenum, and zinc. The results for the use of blast-furnace slag with (N, P and K) seem encouraging regarding the replacement of commercial fertilizers. The use of blast-furnace slag should be enhanced in fertilizer making to reduce the cost of fertilizer manufacturer. The modified slag can be used as fertilizer for agricultural purpose and soil conditioner for acidity corrector of the soil and make it as valuable products to protect environment [11]. Steel slag contains fertilizer components CaO, SiO_2, and MgO [12]. In addition to these three components, it also contains components such as FeO, MnO, and P_2O_5, so it has been used for a broad range of agricultural purposes. Its alkaline property remedies soil acidity. The converter slag is used to produce siliceous fertilizer, phosphorus fertilizer and micronutrient fertilizer [12].

Blast furnace, converter or ladle slags can be used for producing silicate liming materials [13]. Utilization of silicate liming materials neutralizes soil acidity and supplies the soil with plant nutrients. Using blast-furnace lime or converter lime promotes yields, plant quality and soil fertility. The main minerals contain CaO, MgO, SiO_2, Mn and other valuable micro nutrients. The solubility of silicate from slags is often higher than from many other silicate containing soil improvers or rock powders. The basicity of the calcium and magnesium compounds in the slags improves soil pH. The use of steel slags in agriculture produces not only economic but also ecological advantages [13].

The chapter of this book aims to study and to evaluate the antimicrobial efficiency of metallurgical slags for enlarging their construction applications regarding to higher biodegradation resistance. The antimicrobial efficiency of slags was mutually compared as well as compared with a commercial biocide based on antimicrobial silver Ag. The chapter partially originates from the published results in article [14] and contributes to increase the metallurgical solid wastes recovery possibilities and for the utilization of metallurgical slags in ensuring the biological resistance of building materials and products against bio-degradation and bio-deterioration.

2. Materials and methods

The tested kinds of metallurgical slags were as follows: granulated blast-furnace slags with the fineness of 340 m^2/kg (S1a) and 520 m^2/kg (S1b); air cooled blast-furnace slag (S2); demetallized steel slag (S3); calcareous ladle slag (S4); and copper slag (S5). The tested slags were ground to the fineness of 400 m^2/kg, except for S1a and S1b. Their chemical composition was determined by X-ray fluorescence analysis (XRF) using SPECTRO X-LAB 2000 device, according to EN 196-2. The chemical composition of slags is shown in **Table 1**.

The mineralogical composition of the tested slags was analyzed by the XRD method using BRUKER AXS D8 Advance apparatus. The identified mineralogical composition of the slags is as follows: S1a and S1b—glassy phase, melilite C_2AS–C_2MS_2; S2—melilite C_2AS–C_2MS_2, brownmillerite C_4AF, quartz SiO_2; S3—wüstite FeO, brownmillerite C_4AF, free lime CaO, portlandite $Ca(OH)_2$, larnite β-C_2S, quartz SiO_2; S4—free lime CaO, larnite β-C_2S, shanonite γ-C_2S, gehlenite C_2AS, C_3A, gypsum $CaSO_4{\cdot}2H_2O$, quartz SiO_2 and S5—fayalite Fe_2SiO_4, anortite CAS_2, pyroxene type $CaAlAlSiO_6$. Biostat is the commercial biocide additive based on silver Ag bonding to an Al_2O_3 support (Al_2O_3 99.62 wt.%), with the Ag concentration of 20 mg/g. Biostat is added in a proportion of 0.5 wt.% in the thin layer building applications, such as plastering mortars and rendering and of 1.0 wt.% in rough, coarse layer building applications, such as bricklaying and masonry mortar, to achieve an antimicrobial effect.

2.1 Determination of free calcium oxide CaO_{free} content in ground metallurgical slags and pH of slag water leachates

The determination of free calcium oxide CaO_{free} content in the metallurgical slags was performed using the hot ethylene glycol titration method. The slag sample was ground in the laboratory's vibratory mill to a fineness with particle size under 0.1 mm sieve. Then 1.0 g of the slag sample was diluted with 75 mL of ethylene glycol in a filter flask. Afterwards, the solution was heated for half an hour with occasional stirring and then washed with 50 mL of denatured ethanol. Approximately 2–3 drops of alpha-naphthol phthalein were added to the solution as an indicator. The solution was then titrated with 0.1 N HCl, and the titration continued until the color changed from blue to colorless. The CaO_{free} content in the slag sample was calculated as follows: wt.% CaO_{free} = (the volume of 0.1 N HCl used for the titration; in milliliters) × 0.28. The CaO_{free} content in the slag samples is shown in **Table 2**. The ground slag samples were leached in distilled water for 24 h at 20°C. The pH was determined in the slag water leachates by an Agilent Technologies 3200 P pH Meter with an electrode reference system. The measured pH values of the slag water leachates are given in **Table 2**, as well.

No free lime CaO_{free} content was detected in slags S1a, S1b, and S5, very low CaO_{free} content in slags S2 and S3 and the highest CaO_{free} content was measured in slag S4 on the level of 3.30 wt.%, respectively. The pH values of the slag water leachates were in the range of 9.34–12.93. The highest pH values >12 were measured in leachates of slags S3 and S4, caused by the high free lime CaO_{free} content, and the lowest pH < 11 in leachates of slags S2 and S5. The water leached slags of S3 and S4 agglutinated into

Slag	Unit	S1a and S1b	S2	S3	S4	S5
L.O.I[1]	(wt.%)	0.95	0.09	6.02	5.32	+ 4.30[2]
SiO_2	(wt.%)	42.17	40.57	12.81	13.97	27.26
Al_2O_3	(wt.%)	6.87	8.12	1.64	17.77	7.01
Fe_2O_3	(wt.%)	0.32	2.81	29.78	1.90	46.64
CaO	(wt.%)	41.92	41.73	52.30	58.97	7.48
TiO_2	(wt.%)	0.42	0.11	0.34	0.14	0.21
MgO	(wt.%)	10.39	8.44	2.54	3.30	1.90
K_2O	(wt.%)	0.60	0.72	0.04	0.06	0.40
Na_2O	(wt.%)	0.17	0.19	0.07	0.07	1.07
SO_3	(wt.%)	1.84	2.39	0.28	1.98	0.15
MnO	(wt.%)	0.68	2.31	3.54	0.38	0.61
P_2O_5	(wt.%)	0.05	0.14	0.48	0.05	1.26
Cl	(wt.%)	0.0173	0.0112	0.0138	0.0017	0.0012
V	(ppm)	27.0	32.0	298.0	54.0	41.0
Cr	(ppm)	69.6	65.0	981.0	419.0	5740.0
Co	(ppm)	21.1	21.5	98.0	37.6	307.0
Ni	(ppm)	1.9	3.4	9.9	9.8	1893.0
Cu	(ppm)	1.2	1.5	10.1	4.9	7273.0
Zn	(ppm)	98.1	18.7	41.3	12.3	50,341.0
As	(ppm)	0.7	1.4	3.2	22.1	66.65
Cd	(ppm)	11.6	12.0	24.3	22.0	5.0
Sb	(ppm)	1.5	1.8	2.0	26.7	50.2
Hg	(ppm)	2.9	1.9	6.6	4.2	28.5
Tl	(ppm)	1.5	3.3	5.6	6.0	19.0
Pb	(ppm)	4.0	17.6	3.2	7.5	9203.3

[1]*L.O.I: Loss on ignition.*
[2]*Increment on ignition (caused by, e.g., oxidation of sulfides, metallic particles, etc.).*

Table 1.
The chemical composition of the tested slags.

lumpy granules in water due to the hydration of free lime CaO_{free}, β-C_2S, and in the case of slag S4 by the hydration of C_3A with gypsum, as well. The hydrated lumpy granules were mechanically disaggregated to increase the leaching surface contact.

2.2 Determination of antimicrobial efficiency of metallurgical slags under *in vitro* conditions

The antimicrobial activity of the metallurgical slags was tested on selected species of Gram-positive bacteria (G^+), Gram-negative bacteria (G^-), yeasts and filamentous fungi. Microbial strains (bacteria and filamentous fungi) used in the study were either from the Czech Collection of Microorganisms, T. G. Masaryk University, Brno, Czech Republic (CCM) or yeasts from the Collection of Microorganism of the Institute of Biochemistry and Microbiology, Slovak University of Technology, Bratislava, Slovak Republic. Following microorganisms were used: G^+ bacteria—*Bacillus subtilis* CCM 178, *Staphylococcus aureus* CCM 3958,

Slag	Unit	S1a and S1b	S2	S3	S4	S5
CaO free	(wt.%)	0.00	0.06	0.95	3.30	0.00
pH	–	11.71	9.34	12.87	12.93	10.44

Table 2.
The content of free calcium oxide CaO_{free} in the slag samples and pH values of the slag water leachates.

Micrococcus luteus CCM 410; G^- bacteria—*Escherichia coli* CCM 3988, *Pseudomonas aeruginosa* CCM 3630, *Serratia marcescens* CCM 8587; yeasts—*Candida utilis*—*1a*, *Rhodotorula glutinis*—*1*; microscopic filamentous fungi—*Aspergillus niger* CCM-F 384, *Penicillium funiculosum* CCM 8080, *Chaetomium globosum* CCM 8156, *Alternaria alternate* CCM F-128, *Trichoderma viride* CCM F-534, *Cladosporium herbarum* CCM F-534.

2.2.1 Growth media

1. Meat-peptone bouillon (broth) for inoculation of bacteria (containing in 1000 mL of distilled water, 5 g peptone for bacteriology, 5 g meat extract and 2.5 g NaCl, with the pH adjusted from 7.2 to 7.4);

2. Meat-peptone agar for cultivation of bacteria (containing in 1000 mL of distilled water, 5 g peptone for bacteriology, 5 g meat extract, 2.5 g NaCl and 20 g agar, with the pH adjusted from 7.2 to 7.4);

3. Sabouraud's glucose bouillon (broth) for inoculation of yeasts (containing in 1000 mL of distilled water, 10 g peptone for bacteriology and 20 g glucose, with the pH adjusted to 6.5);

4. Malt agar for cultivation of yeasts and filamentous fungi (containing in 1000 mL of distilled water, 20 g agar and 60 g malt extract, with the pH adjusted to 6.5).

2.2.2 Solutions

For diluting the inoculum of bacteria and yeasts, saline solution (0.85% NaCl) was used, and for preparation of the spore suspension (filamentous fungi), a 0.1% water solution of Tween 80 was used.

2.2.3 Preparation of the inoculum of bacteria and yeasts

A total of 25 mL of meat-peptone bouillon (broth) in a 100 mL Erlenmeyer flask was inoculated with a three-day-old culture of model bacteria using a bacterial loop, and 25 mL of Sabouraud's glucose bouillon (broth) in a 100 mL Erlenmeyer flask was inoculated with a three-day-old culture of yeasts using an inoculation loop. The microorganism cultures were incubated for 15 h at 30°C using shaking apparatus (at a vibration frequency 4 Hz). The grown microorganism cultures were aseptically filtered through three-ply gauze in order to remove possible clusters of cells. The obtained cell suspensions were diluted 100-fold with sterile saline (aqueous physiological solution) and were used for inoculation of the growth media (the cells concentration was 10^6 cells per mL) in order to determine the antibacterial and anti-yeast activity of the slag samples, respectively.

2.2.4 Preparation of spore suspension of filamentous fungi

A total of 8 mL 0.1% water solution of Tween 80 was aseptically added to the sporulated cultures of filamentous fungi, which had been cultivated on slant malt agar for 21 days. Spores from mycelium were loosened by bacterial loop, and filtered as above. The filtered spore suspension was diluted by sterile saline (the spores concentration was 2×10^7 spores per mL) and used for inoculation of the growth media in order to determine the antifungal activity of the slag samples.

2.2.5 Determination of antimicrobial activity

Meat-peptone bouillon for inoculation of bacteria, meat-peptone agar for cultivation of bacteria, Sabouraud's glucose bouillon for inoculation of yeasts and malt agar for cultivation of yeasts and filamentous fungi were used as growth media. Saline solution was used for diluting the inoculum of bacteria and yeasts, and a water solution of Tween 80 was used for preparation of the spore suspension. Sterilization of growth media and solutions was realized in the laboratory autoclave at 120°C for 20 min. The tests were performed in the laboratory incubator at temperatures of 30°C for bacteria, 28°C for yeasts and 25°C for filamentous fungi at a relative humidity of 95%. Antimicrobial activity was determined by dilution methods in agar growth media, so that the resulting concentration of tested slags in growth media was 10, 20, 40, and 60%; further, the concentration of Biostat in growth media was 1.0 and 0.5%, respectively. The pH of the growth media with the addition of the slags was strongly alkaline (pH 11); thus, half the samples of each slag was tested at this pH, and the second half of the samples was tested at a modified pH (bacteria pH 7.2, yeasts, and filamentous fungi pH 6.6). The first half of the samples with the original pH represented real conditions for growth of microorganisms in concrete; the second half of the samples with a modified pH represented optimal conditions for growth of the microorganisms *in vitro*.

After sterilization, meat-peptone agar and malt agar with added slag samples and Biostat (at original pH and modified pH) were cooled down to 60°C. The slag samples and Biostat were equally dispersed throughout growth media and subsequently they were divided and 6 mL of each was placed in the Petri dishes Ø 60.0 mm. Sterile paper discs (Ø 5.0 mm) were placed on the surface of solidified growth media, which were inoculated by 5 μL from spore suspension of model filamentous fungi on malt agar. The inoculation of bacteria and yeasts was done by direct pipetting the suspension on the surface of agarized media. Each 10 μL of suspensions were pipetted from the model bacteria onto meat-peptone agar or model yeasts on the malt agar.

The microorganisms were incubated in a thermostat at temperatures of 30°C for bacteria, 28°C for yeasts, 25°C for filamentous fungi at a relative humidity of 95% for four days. The growth intensity of bacteria and yeasts was compared with the growth of bacteria and yeasts in the control growth media without slags. The growth of filamentous fungi in the presence of slags was monitored by the measurement of average diameter of growing colony at regular time intervals and was compared with the filamentous fungi growth in the control growth media without slags. In the case that no filamentous fungi growth with the presence of slags was observed, the paper discs with spores were transferred on fresh growth media. After 96 h incubation at 25°C, the inhibiting effect of slags was inspected: fungistatic (spores germinate and subsequently the mycelium grows broader) and fungicidal-lethal (fungi do not grow; they are dead).

3. Results

The antimicrobial efficiency of metallurgical slags on selected species of G^+ bacteria, G^- bacteria, yeasts and filamentous fungi was determined and then mutually compared. The antimicrobial efficiency results are given in **Tables 3–8**. The antibacterial efficiencies of slags are given in **Tables 3** and **4**; the anti-yeast efficiencies are shown in **Table 5**; and the antifungal efficiencies are given in **Tables 6–8**, respectively. The antimicrobial efficiencies of slags are reciprocally compared as well as compared with the antimicrobial efficiencies of the commercial biocide additive—Biostat. The growth extent of bacteria and yeasts is expressed as the signs (−, +, ++, +++), and the growth degree of filamentous fungi is expressed as the percentage (%) of sample surface covered by fungi colonies (measured against a K—control growth media without slags; with the expression of s—fungistatic effect and c—fungicidal effect). The tests were carried out at N—neutral pH of growth media and A—alkaline pH of growth media, as well.

S1aA	*M. luteus*	*S. aureus*	*B. subtilis*	*E. coli*	*S. marcescens*	*P. aeruginosa*
10%	+++	+++	+++	+++	+++	+++
20%	+++	+++	+++	+++	+++	+++
40%	+++	++	−	++	+++white	+++
60%	+++	++	−	+	++white	+++
S1aN	***M. luteus***	***S. aureus***	***B. subtilis***	***E. coli***	***S. marcescens***	***P. aeruginosa***
10%	+++	+++	+++	+++	+++	+++
20%	+++	+++	+++	+++	+++	+++
40%	+++	+++	−	+++	+++	+++
60%	+++	−	−	+++	+++	+++
S1bA	***M. luteus***	***S. aureus***	***B. subtilis***	***E. coli***	***S. marcescens***	***P. aeruginosa***
10%	+++	+++	+++	+++	+++white	+++
20%	+++	+++	+++	+++	+++white	+++
40%	+++	+++	−	++	+++white	+++
60%	+++	+++	−	+	++white	+++
S1bN	***M. luteus***	***S. aureus***	***B. subtilis***	***E. coli***	***S. marcescens***	***P. aeruginosa***
10%	+++	+++	+++	+++	+++	+++
20%	+++	+++	+++	+++	+++	+++
40%	+++	+++	−	+++	+++	+++
60%	+++	++	−	−	+++	+++
S2A	***M. luteus***	***S. aureus***	***B. subtilis***	***E. coli***	***S. marcescens***	***P. aeruginosa***
10%	++	+	+	++	++white	++
20%	++	−	−	++	++white	++
40%	++	−	−	++	++white	++
60%	−	−	−	−	−	−
S2N	***M. luteus***	***S. aureus***	***B. subtilis***	***E. coli***	***S. marcescens***	***P. aeruginosa***
10%	+++	+++	−	+++	+++	+++
20%	+++	+++	−	+++	+++	+++

40%	+++	+++	–	+++	+++	+++
60%	+++	+++	–	++	++	++
S3A	***M. luteus***	***S. aureus***	***B. subtilis***	***E. coli***	***S. marcescens***	***P. aeruginosa***
10%	+++	+	+	+++	+++	+++
20%	+++	–	–	+++	+++	+++
40%	+++	–	–	++	+++	+++
60%	+++	–	–	+	+++	+++

Notes: –, no growth of bacteria; +, the growth of bacteria is negligible; ++, the growth of bacteria is gradual; +++, the growth of bacteria is intensive comparable with the control growth media; A, alkaline pH of growth media; N, neutral pH of growth media; white, the colony of Serratia marcescens—loss of red pigment by slag impact.

Table 3.
Antibacterial efficiency of slags regarding bacterial growth inhibition.

3.1 Antibacterial efficiency of metallurgical slags

The test results of antibacterial efficiency of slags and Biostat on the selected model G^+ bacteria and G^- bacteria are shown in **Tables 3** and **4**. From the results, it is evident that the inhibitory effect of slags differs. Slag S4 had the highest antibacterial activity and intensely inhibited growth of G^+ and also G^- bacteria. It was proven already at the lowest concentration of slag S4: 10% in growth media. The bacteria, except *M. luteus*, did not grow at higher slag S4 concentrations, whereas the pH values of growth media (neutral, alkaline) did not affect the intensity of bacterial growth inhibition.

The slag S3 inhibited the growth of G^+ bacteria *S. aureus* and *B. subtilis* already at the concentration of 10%, and at higher slag S3 concentrations the growth was completely inhibited. However, slag S3 did not affect the growth of G^- bacteria, except for *E. coli*, at the highest concentration of 60%.

Bacterial growth inhibition with slag S2 was more intensive in the alkaline pH of growth media. The complete inhibition of the growth of G^+ bacteria *B. subtilis* and *S. aureus* was measured at slag S2 concentrations of 20, 40, and 60% and, inhibition of the growth of G^- bacteria *S. marcescens, P. aeruginosa* and *E. coli* was detected at the slag S2 concentration of 60%. Slag S2 only completely inhibited the growth of *B. subtilis* at a neutral pH of growth media and at all tested concentrations. Slag S1a caused total inhibition of *B. subtilis* growth at concentrations of 40 and 60% at both a neutral and alkaline pH of growth media. The growth of *S. aureus* was completely inhibited at slag S1a concentration of 60% at a neutral pH of growth media. The 40 and 60% slag S1b concentration at neutral as well as alkaline pH of growth media caused the total inhibition of growth of *B. subtilis*. No growth of *E. coli* was measured at slag S1b concentration of 60% at neutral pH of growth media. Slag S5 did not significantly affect the growth of bacteria. Slags S1a, S1b, S2 and S5 brought about a change in *S. marcescens* growth at alkaline pH. The bacteria *S. marcescens* lost its red pigment and grew as a white colony. Biostat at concentration of 1.0% caused 100% growth inhibition of all bacteria at neutral and alkaline pH of growth media. Only negligible or no bacterial growth was observed at Biostat concentration of 0.5% in growth media, and pH had no effect on their growth. Based on the measured results, it is evident that the antibacterial efficiency of slags decreased in the order: S4 > S3 > S2 > S1a = S1b > S5. The testing arrangement in Petri dishes regarding the bacterial growth inhibition with slag S2 at neutral pH (N) and alkaline pH (A) of growth media is given in **Figure 1A**.

S3N	*M. luteus*	*S. aureus*	*B. subtilis*	*E. coli*	*S. marcescens*	*P. aeruginosa*
10%	+++	++	++	+++	+++	+++
20%	+++	–	–	+++	+++	+++
40%	+++	–	–	+++	+++	+++
60%	+++	–	–	–	+++	+++
S4A	***M. luteus***	***S. aureus***	***B. subtilis***	***E. coli***	***S. marcescens***	***P. aeruginosa***
10%	++	++	++	++	+	–
20%	+	–	–	–	–	–
40%	+	–	–	–	–	–
60%	–	–	–	–	–	–
S4N	***M. luteus***	***S. aureus***	***B. subtilis***	***E. coli***	***S. marcescens***	***P. aeruginosa***
10%	+	+	+	+	+	–
20%	+	–	–	–	–	–
40%	+	–	–	–	–	–
60%	–	–	–	–	–	–
S5A	***M. luteus***	***S. aureus***	***B. subtilis***	***E. coli***	***S. marcescens***	***P. aeruginosa***
10%	+++	+++	+++	+++	+++	+++
20%	+++	+++	+++	+++	+++	+++
40%	+++	+++	+	+++	+++white	+++
60%	+++	+++	+	+++	++white	+++
S5N	***M. luteus***	***S. aureus***	***B. subtilis***	***E. coli***	***S. marcescens***	***P. aeruginosa***
10%	+++	+++	+++	+++	+++	+++
20%	+++	+++	+++	+++	+++	+++
40%	+++	+++	+++	+++	+++	+++
60%	++	++	++	++	++	++
BiostatA	***M. luteus***	***S. aureus***	***B. subtilis***	***E. coli***	***S. marcescens***	***P. aeruginosa***
0.5%	++	–	+	+	+	+
1%	–	–	–	–	–	–
BiostatN	***M. luteus***	***S. aureus***	***B. subtilis***	***E. coli***	***S. marcescens***	***P. aeruginosa***
0.5%	+	–	–	–	+	–
1%	–	–	–	–	–	–

Notes: –, no growth of bacteria; +, the growth of bacteria is negligible; ++, the growth of bacteria is gradual; +++, the growth of bacteria is intensive comparable with the control growth media; A, alkaline pH of growth media; N, neutral pH of growth media; white, the colony of Serratia marcescens—loss of red pigment by slag impact.

Table 4.
Antibacterial efficiency of slags regarding bacterial growth inhibition.

3.2 Anti-yeast efficiency of metallurgical slags

The test results of anti-yeast efficiency of slags and Biostat on the selected model yeasts are shown in **Table 5**. Slag S4 had the highest anti-yeast activity. Growth of all yeasts was totally inhibited at slag S4 concentration as low as 10%, at both alkaline and neutral pH of growth media. Slags S1a, S1b, and S3 caused complete

S1aA	C. utilis	R. glutinis	S1aN	C. utilis	R. glutinis
10%	+++	+++	10%	+++	+++
20%	–	–	20%	–	–
40%	–	–	40%	–	–
60%	–	–	60%	–	–
S1bA	***C. utilis***	***R. glutinis***	**S1bN**	***C. utilis***	***R. glutinis***
10%	+	+	10%	+	+
20%	–	–	20%	–	–
40%	–	–	40%	–	–
60%	–	–	60%	–	–
S2A	***C. utilis***	***R. glutinis***	**S2N**	***C. utilis***	***R. glutinis***
10%	+++	+++	10%	+++	+++
20%	+++	++	20%	+++	+++
40%	–	+	40%	+	+
60%	–	+	60%	–	+
S3A	***C. utilis***	***R. glutinis***	**S3N**	***C. utilis***	***R. glutinis***
10%	–	+	10%	–	+++
20%	–	–	20%	–	–
40%	–	–	40%	–	–
60%	–	–	60%	–	–
S4A	***C. utilis***	***R. glutinis***	**S4N**	***C. utilis***	***R. glutinis***
10%	–	–	10%	–	–
20%	–	–	20%	–	–
40%	–	–	40%	–	–
60%	–	–	60%	–	–
S5A	***C. utilis***	***R. glutinis***	**S5N**	***C. utilis***	***R. glutinis***
10%	+++	+++	10%	+++	+++
20%	++	++	20%	++	+++
40%	++	++	40%	++	+++
60%	+	+	60%	++	++
BiostatA	***C. utilis***	***R. glutinis***	**BiostatN**	***C. utilis***	***R. glutinis***
0.5%	+++	+++	0.5%	+++	+++
1%	–	–	1%	–	–

Notes: –, no yeasts growth; +, yeasts growth is negligible; ++, yeasts growth is gradual; +++, yeasts growth is intensive comparable with the control growth media; N, neutral pH of growth media; A, alkaline pH of growth media.

Table 5.
Growth inhibition of selected yeasts by metallurgical slags.

growth inhibition of all yeasts from a concentration of 20% at neutral and alkaline pH of growth media, as well. The growth of yeasts was intensively reduced with slag S2 at concentrations of 40 and 60%. Slag S5 partially inhibited the growth of yeasts; however, total growth inhibition was not observed even at the highest slag S5 concentration of 60%. Biostat at concentration of 1.0% caused total growth inhibition of all yeasts at both neutral, as well as alkaline, pH of growth media. However,

Sample	Growth of selected filamentous fungi (%)					
S1aA	***A. alternata***	***P. funiculosum***	***A. niger***	***T. viride***	***C. herbarum***	***Ch. globosum***
10%	90	30	0 s	0 s	88	72
20%	78	24	0 s	0 s	80	20
40%	50	0s	0 s	0 s	60	0 s
60%	50	0s	0 s	0 s	40	0 c
K	100	100	100	100	100	100
S1aN	***A.alternata***	***P.funiculosum***	***A.niger***	***T. viride***	***C.herbarum***	***Ch. globosum***
10%	90	30	0 s	0 s	88	72
20%	60	24	0 s	0 s	80	20
40%	30	0s	0 s	0 s	20	0 s
60%	0s	0s	0 s	0 s	0 s	0 c
K	100	100	100	100	100	100
S1bA	***A.alternata***	***P.funiculosum***	***A.niger***	***T. viride***	***C.herbarum***	***Ch. globosum***
10%	80	22	0 s	0 s	80	72
20%	60	14	0 s	0 s	40	60
40%	30	10	0 s	0 s	40	50
60%	0s	0	0 s	0 s	0 s	10
K	100	100	100	100	100	100
S1bN	***A.alternata***	***P.funiculosum***	***A.niger***	***T. viride***	***C.herbarum***	***Ch. globosum***
10%	40	30	0 s	0 s	80	72
20%	40	14	0 s	0 s	60	60
40%	40	10	0 s	0 s	60	40
60%	0 s	0 s	0 s	0 s	0 s	10
K	100	100	100	100	100	100
S2A	***A.alternata***	***P.funiculosum***	***A.niger***	***T. viride***	***C.herbarum***	***Ch. globosum***
10%	80	30	0 s	0 s	80	80
20%	70	26	0 s	0 s	80	60
40%	54	20	0 s	0 s	60	40
60%	52	15	0s	0s	60	40
K	100	100	100	100	100	100

Notes: s, fungistatic effect; c, fungicidal effect; N, neutral pH of growth media; A, alkaline pH of growth media; K, control growth media.

Table 6.
Antifungal efficiency of metallurgical slags regarding inhibition of growth of selected filamentous fungi.

the intensive growth of yeasts was measured at Biostat concentration of 0.5%, and so Biostat did not possess an inhibiting effect on the growth of yeasts at this concentration. Based on the obtained results, it is evident that the anti-yeast efficiency of slags decreased in the order: S4 > S1a = S1b = S3 > S2 > S5. The testing arrangement in Petri dishes regarding the yeasts growth inhibition with slag S5 at alkaline pH (A) and neutral pH (N) of growth media is given in **Figure 1B**.

3.3 Antifungal efficiency of metallurgical slags

The test results of antifungal efficiency of slags and Biostat on the selected model filamentous fungi are shown in **Tables 6–8**. Growth of *A. alternate* was most intensively inhibited in the presence of slag S4. Total inhibition of *A. alternata* growth was measured at slag S4 concentrations of 40 and 60% at alkaline as well as neutral pH of growth media, particularly at concentration of 20% at alkaline pH. Slag S4 even possessed a fungicidal (lethal) effect on the fungal spores of *A. alternata* at concentration of 60% at neutral pH, while slag S1b hindered the growth of *A. alternata* at concentration of 60% at alkaline as well as neutral pH of growth media. Slags S1a and S3 totally inhibited *A. alternata* growth at concentration of 60%, but only at neutral pH of the growth media. Slag S2 at concentrations of 40 and 60%, and slag S5 at concentration of 60%, partially inhibited the growth of *A.*

Sample	**Growth of selected filamentous fungi (%)**					
S2N	***A. alternata***	***P. funiculosum***	***A. niger***	***T. viride***	***C. herbarum***	***Ch. globosum***
10%	80	40	0 s	0 s	100	70
20%	72	30	0 s	0 s	80	60
40%	64	24	0 s	0 s	70	60
60%	54	20	0 s	0 s	60	50
K	100	100	100	100	100	100
S3A	***A. alternata***	***P. funiculosum***	***A. niger***	***T. viride***	***C. herbarum***	***Ch. globosum***
10%	80	20	0 s	0 s	52	80
20%	70	10	0 s	0 s	40	75
40%	40	0s	0 s	0 s	20	60
60%	20	0s	0 s	0 s	0 s	0 s
K	100	100	100	100	100	100
S3N	***A. alternata***	***P. funiculosum***	***A. niger***	***T. viride***	***C. herbarum***	***Ch. globosum***
10%	60	10	0 s	0 s	60	80
20%	40	0 s	0 s	0 s	40	70
40%	32	0 s	0 s	0 s	28	60
60%	0 s	0 s	0 s	0 s	0 s	0 s
K	100	100	100	100	100	100
S4A	***A. alternata***	***P. funiculosum***	***A. niger***	***T. viride***	***C. herbarum***	***Ch. globosum***
10%	70	0 s	0 s	0 s	60	0 s
20%	0 s	0 s	0 s	0 s	0 s	0 s
40%	0 s	0 s	0 s	0 s	0 s	0 s
60%	0 s	0 s	0 s	0 c	0 s	0 c
K	100	100	100	100	100	100

Sample	Growth of selected filamentous fungi (%)					
S4N	*A. alternata*	*P. funiculosum*	*A. niger*	*T. viride*	*C. herbarum*	*Ch. globosum*
10%	20	20	0 s	0 s	0 s	0 s
20%	10	10	0 s	0 s	0 s	0 s
40%	0 s	0 s	0 s	0 s	0 s	0 s
60%	0 c	0 s	0 c	0 c	0 s	0 c
K	100	100	100	100	100	100

Notes: s, fungistatic effect; c, fungicidal effect; N, neutral pH of growth media; A, alkaline pH of growth media; K, control growth media.

Table 7.
Antifungal efficiency of metallurgical slags regarding inhibition of growth of selected filamentous fungi.

alternata. Based on the comparison of growth intensity of *A. alternata* in the presence of the tested slags, the order of decreasing inhibition efficiency of slags was: S4 > S1b > S1a = S3 > S2 > S5.

Growth of *P. funiculosum* was significantly inhibited by all tested slags. The highest growth inhibition was measured in the presence of slag S4. Based on a comparison of the growth intensity of *P. funiculosum* in the presence of tested slags, the order of inhibiting efficiency of slags decreased as follows: S4 > S3 > S1a > S1b > S5 > S2. Total growth inhibition of *A. niger* was measured in the presence of all tested slags, with the exception of slag S5, at concentration of 10% at both alkaline and neutral pH of growth media. Slag S4 had a fungicidal (lethal) effect on the fungal spores of *A. niger* at concentration of 60% at neutral pH of growth media. The order of inhibiting efficiency of tested slags on *A. niger* growth decreased as follows: S4 > S1a = S1b = S2 = S3 > S5. No *T. viride* growth was measured in the presence of all tested slags, with exception of slag S5, at concentration of 10% at alkaline pH of growth media. Slag S4 possessed a fungicidal (lethal) effect on fungal spores of *T. viride* at concentration of 60% at neutral as well as alkaline pH of growth media. The inhibiting efficiency of the tested slags on *T. viride* growth decreased in the order: S4 > S1a = S1b = S2 = S3 > S5.

Slag S4 had the highest inhibiting efficiency on *C. herbarum* growth. Complete growth inhibition of *C. herbarum* was detected at all tested slag S4 concentrations, with the exception of concentration of 10% at alkaline pH of growth media, when the growth inhibition of *C. herbarum* was on the level of 40%. No growth was detected in the presence of S1b and S3 slags at concentration of 60% at either alkaline or neutral pH and in the presence of S1a at neutral pH of growth media. Partial growth inhibition of *C. herbarum* was measured at slags S2 and S5. The comparison of growth intensity of *C. herbarum* in the presence of the tested slags demonstrated that the inhibiting efficiency of the tested slags decreased in the order: S4 > S3 > S1b > S1a > S5 > S2. Total growth inhibition of *Ch. globosum* was measured at slag S4, even from the lowest concentration of 10%. Similarly, complete growth inhibition was observed in the presence of slags S1a and S5, however, only at high concentration values of 40 and 60%, and as in the presence of slag S3 even at the highest concentration of 60%. Partial growth inhibition of *Ch. globosum* was measured at slags S1b and S2. The comparison of growth intensity of *Ch. globosum* in the presence of the tested slags showed a decreasing inhibition efficiency of slags in the order: S4 > S1a = S5 > S3 > S1b > S2.

From the measured results, it is seen that the model filamentous fungi used were sensitive to the presence of tested slags in various ways. All slags inhibited the growth of filamentous fungi by 40–100% at concentration of 60% slag. The most sensitive to the presence of all slags were *T. viride* and *A. niger*. Their growth was completely inhibited at 20–60% concentrations of all slags at alkaline and neutral pH of growth media. The growth of these two fungi species was stopped, thus causing a fungistatic effect. The 60% concentration of slag S4 resulted in the inhibition of *T. viride* growth with fungicide effect (killed fungal spores). Slag S4 possessed the most inhibiting activity for all filamentous fungi. It totally inhibited growth of almost all filamentous fungi in concentrations from 20 to 60% at alkaline pH of growth media, with mostly a fungistatic effect. Only *T. viride* growth was inhibited completely at slag S4 concentration of 60% at alkaline pH of growth media with a fungicide effect. The growth of filamentous fungi was inhibited completely at concentrations from 40 to 60% of slag S4 and at slag S4 concentration of 60% with fungicide effect on *A. niger, A. alternate, Ch. globosum* and *T. viride*, at neutral pH of growth media.

A lower inhibiting effect on filamentous fungi growth was measured at slags S1a, S1b and S3, however, they inhibited the growth of all filamentous fungi from

Sample	**Growth of selected filamentous fungi (%)**					
S5A	***A. alternata***	***P. funiculosum***	***A. niger***	***T. viride***	***C. herbarum***	***Ch. globosum***
10%	84	48	20	20	60	60
20%	80	14	0 s	0 s	40	10
40%	80	10	0 s	0 s	40	0 s
60%	56	0	0 s	0 s	40	0 s
K	100	100	100	100	100	100
S5N	***A. alternata***	***P.funiculosum***	***A. niger***	***T. viride***	***C.herbarum***	***Ch.globosum***
10%	80	52	70	0 s	72	80
20%	80	20	0 s	0 s	40	40
40%	80	10	0 s	0 s	40	0 s
60%	50	10	0 s	0 s	40	0 s
K	100	100	100	100	100	100
BiostatA	***A. alternata***	***P. funiculosum***	***A. niger***	***T. viride***	***C. herbarum***	***Ch. globosum***
0.5%	100	100	100	100	100	60
1%	10	70	70	40	50	30
K	100	100	100	100	100	100
BiostatN	***A. alternata***	***P. funiculosum***	***A. niger***	***T. viride***	***C. herbarum***	***Ch. globosum***
0.5%	100	100	100	100	100	60
1%	30	70	70	40	50	30
K	100	100	100	100	100	100

Notes: s, fungistatic effect; c, fungicidal effect; N, neutral pH of growth media; A, alkaline pH of growth media; K, control growth media.

Table 8.
Antifungal efficiency of metallurgical slags regarding inhibition of growth of selected filamentous fungi.

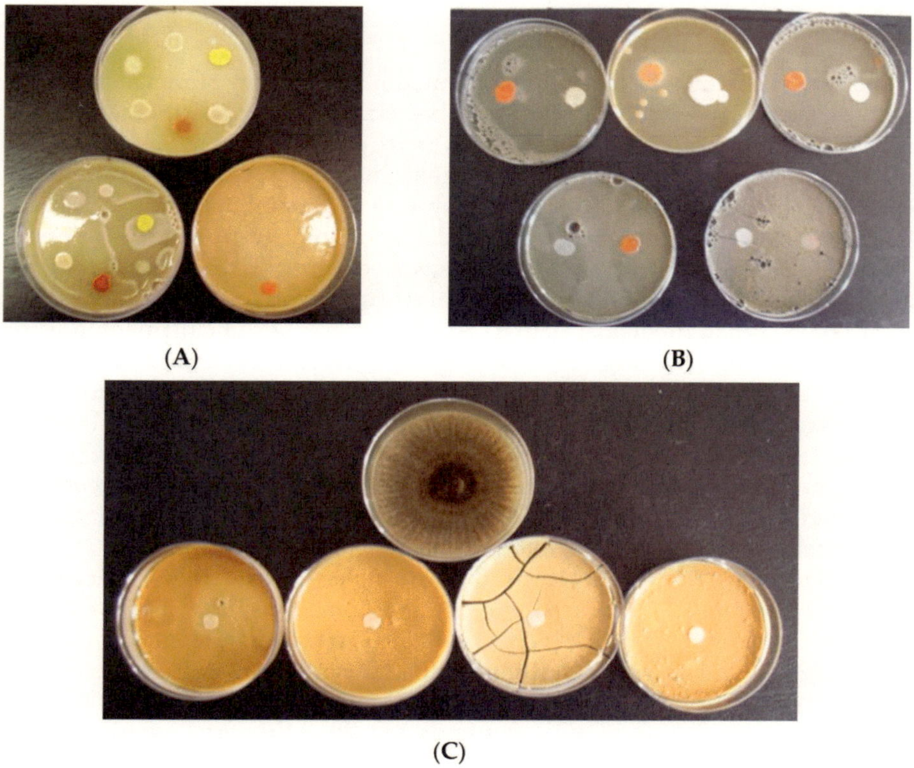

Figure 1.
*The testing arrangement in petri dishes regarding the growth inhibition of selected species of G^+ bacteria, G^- bacteria, yeasts and filamentous fungi in the presence of tested slags: (**A**) bacterial growth inhibition with slag S2 at neutral pH (N) and alkaline pH (A) of growth media (upper line: control (no slag); lower line from left: slag S2 (N)—60%, S2 (A)—60%); (**B**) yeasts growth inhibition with slag S5 at alkaline pH (A) and neutral pH (N) of growth media (upper line from left: slag S5 (A)—10%, control (no slag), slag S5 (N)—10%; lower line from left: slag S5 (A)—60%, S5 (N)—60%); and (**C**) inhibition of Aspergillus niger growth with slag S1a at alkaline pH (A) of growth media (upper line: control (no slag), lower line from left: slag S1a (A)—10, 20, 40, 60%).*

40 to 100% at concentration of 40%. The lowest impact on the growth of filamentous fungi was observed at slags S2 and S5, which affected only the growth of the most sensitive fungi, *A. niger* and *T. viride*. The most resistant was *A. alternata*. Its growth was most intensely inhibited by slag S4, with total inhibition observed at 40–60% of slag concentration range, with mostly a fungistatic effect. Biostat, at concentration of 0.5% did not affect the growth of filamentous fungi, with exception *Ch. globosum*, the growth of which was inhibited by 40%. Biostat at concentration of 1.0% inhibited the growth of filamentous fungi by 30–70%. Based on the measured results, it is evident that Biostat is characterized mainly by antibacterial and anti-yeast efficiency, and affects the growth of model filamentous fungi only to a minimum extent. Regarding the fact that model filamentous fungi were selectively sensitive to the presence of the tested slags, it is possible to determine only an approximate order of inhibition efficiency of the slags to filamentous fungi as follows: S4 > S3 = S1a = S1b > S5 > S2. The pH values of the growth media did not significantly influence the intensity of inhibition of the model microorganisms' growth. The testing arrangement in Petri dishes regarding the inhibition of *A. niger* growth with slag S1a in alkaline pH (A) of growth media is given in **Figure 1C**.

4. Discussion

The results obtained by testing the antimicrobial efficiency of metallurgical slags on selected species of G^+ bacteria, G^- bacteria, yeasts and filamentous fungi can be summarized as follows: the tested slags are characterized by selective toxicity on the model representatives of biodeteriogenic microflora; calcareous ladle slag (S4) has the highest antimicrobial efficiency; slag S4 best inhibits the growth of bacteria, yeasts, and filamentous fungi of all slags; complete growth inhibition of *A. niger* and *T. viride* was measured at all slags at concentration of 10%, except for copper slag (S5); slag (S4) and granulated blast-furnace slag with the fineness of 340 m^2/kg (S1a) had fungicidal (lethal) effect on spores of some filamentous fungi at concentration of 60%; slag (S5) possesses the lowest antibacterial and anti-yeast efficiency, despite having the highest Cu, Hg, As, Pb, Cr, Zn, Co, and Ni amounts from all slags; air-cooled blast-furnace slag (S2) and slag (S5) possess the lowest antifungal efficiency. Biostat at concentrations of 0.5 and 1% inhibited the growth of bacteria, at concentration of 1% inhibited the growth of yeasts, at concentration of 0.5% did not affect the growth of filamentous fungi at all, and at concentration of 1% inhibited the growth of filamentous fungi only to a minimum extent.

In general, slag (S4) is characterized by the highest antimicrobial efficiency and inhibits the growth of bacteria, yeasts and filamentous fungi most of all slags. Slag S4 possesses the highest CaO content of 58.97 wt.% and the highest CaO_{free} content (3.30 wt.%) as well as the highest pH value of water leachate (12.93), in comparison with the other slags. Antimicrobial activity of metallurgical slags, with the very high CaO and CaO_{free} contents, as well as the very high pH value of water leachate, as in the case of slag S4, could be approximately similar to the antimicrobial efficiency of lime [15, 16]. Calcareous metallurgical slags give rise to an increase in the pH values of water leachates up to the range of about 11.5–12.9 due to the high CaO content, e.g., similarly in the case of granulated blast-furnace slag, cements and concretes in the wet state, when they hydrate [1, 2, 6], which affects the viability of microorganisms. This effect is similar to the effect of lime during hygienisation. Disinfection by lime reduces extensively the number of microorganisms when operated at pH 11.2 at a wastewater treatment plant [15]. Lime enables the destruction of all pathogens due to the pH increase (alkaline hydrolysis), combined with the temperature rise that quicklime hydration provides (thermolysis) [16]. The thermolysis effect can be excluded at metallurgical slags in comparison with lime, however, alkaline hydrolysis could be considered as their antimicrobial potential. Alkaline hydrolysis (OH^-) results in the destruction of proteins, nucleic acids (RNA), carbohydrate and lipids and so finally pathogen inactivation [16].

5. Conclusions

According to the measured results, it is evident that: the antibacterial efficiency of the tested metallurgical slags decreased in the order: S4 > S3 > S2 > S1a = S1b > S5; the decrease in anti-yeast efficiency was in the order: S4 > S1a = S1b = S3 > S2 > S5; filamentous fungi were selectively sensitive to the tested slags, therefore, it is only possible to determine the approximate order of inhibition efficiency of slags to filamentous fungi: S4 > S3 = S1a = S1b > S5 > S2. Calcareous ladle slag (S4) is characterized by the highest antimicrobial efficiency, while granulated blast-furnace slag (S1a, S1b) and demetallized steel slag (S3) possess medium activity. Air cooled blast-furnace slag (S2) has still lower activity and copper slag (S5) has the lowest activity. Based on the antimicrobial efficiency results, metallurgical slags possess a great potential for utilization as appropriate antimicrobial components for construction applications.

Acknowledgements

The chapter is partially based on the results and information from the published article [14] "Strigáč J, Števulová N, Mikušinec J, Varečka Ľ, Hudecová D. Antimicrobial efficiency of metallurgical slags for application in building materials and products. Buildings 2018;8:33. DOI: 10.3390/buildings8020033," which was published by MDPI AG, Buildings Editorial Office, Basel, Switzerland (http://www.mdpi.com/journal/buildings/). MDPI AG, Buildings Editorial Office in the published article [14] contributed to the chapter of this book. The chapter is extended review paper of the article [14], with new insights.

Notes/thanks/other declarations

Many thanks to the original publisher—MDPI AG, Buildings Editorial Office, St. Alban-Anlage 66, 4052 Basel, Switzerland (http://www.mdpi.com/journal/buildings/) of article [14].

Author details

Július Strigáč[1*], Nadežda Števulová[2], Jozef Mikušinec[1], Ľudovít Varečka[3] and Daniela Hudecová[3]

1 Povazska Cement Plant Ladce, Ladce, Slovakia

2 Institute of Environmental Engineering, Faculty of Civil Engineering, Technical University of Košice, Košice, Slovakia

3 Institute of Biochemistry and Microbiology, Faculty of Chemical and Food Technology, Slovak University of Technology in Bratislava, Bratislava, Slovakia

*Address all correspondence to: strigac.j@pcla.sk

References

[1] Cwalina B. Biodeterioration of concrete. Architecture Civil Engineering Environment. 2008;**4**:133-140

[2] George RP, Vishwakarma V, Samal SS, Mudali UK. Current understanding and future approaches for controlling microbially influenced concrete corrosion: A review. Concrete Research Letters. 2012;**3**:491-506

[3] House MW, Weiss WJ. Review of microbially induced corrosion and comments on needs related to testing procedures. In: Olek J, Weiss J, editors. Proceedings of the 4th International Conference on the Durability of Concrete Structures(ICDCS), 24-26 July 2014; West Lafayette, IN, USA: Purdue University; 2014. pp. 94-103. Available from: https://pdfs.semanticscholar.org/38e9/5fe75db5b0043d05b7615b76221842bbe19d.pdf

[4] Strigáč J, Martauz P, Eštoková A, Števulová N, Luptáková A. Bio-corrosion resistance of concretes containing antimicrobial ground granulated blastfurnace slag BIOLANOVA and novel hybrid H-CEMENT. Solid State Phenomena. 2016;**244**:57-64. DOI: 10.4028/www.scientific.net/SSP.244.57

[5] Giannantonio DJ, Kurth JC, Kurtis KE, Sobecky PA. Effects of concrete properties and nutrients on fungal colonization and fouling. International Biodeterioration and Biodegradation. 2009;**63**:252-259. DOI: 10.1016/j.ibiod.2008.10.002

[6] Strigáč J, Martauz P. Fungistatic properties of granulated blastfurnace slag and related slag-containing cements. Ceramics–Silikáty. 2016;**60**:19-26. DOI: 10.13168/cs.2016.0003

[7] Anjaneyulu E, Ramgopal M, Narasimha G, Balaji M. Effect of pig iron slag particles on soil physico-chemical, biological and enzyme activities. Iranica Journal of Energy and Environment. 2011;**2**:161-165

[8] Papastergiadis E, Nathanail E, Sampalioti D, Sklari S, Chasiotis A, Lollos K, Karagiannidis A, Samaras P. The use of steelmaking slag for sewage sludge stabilization. In: Albanis T et al. editors. Proceedings of the 10th International Conference on Protection and Restoration of the Environment (PRE 10), 5-9 July 2010; Corfu Island, Greece. 2010;**191**:5-9. Available from: http://ikee.lib.auth.gr/record/251683; https://qa.auth.gr/en/cv/akarag?nopagecache=1

[9] Stanojković A, Pivić R, Maksimović S, Stevanović D, Delić D, Stajković O. The effect of metallurgical slag on microbiological activity in Pseudogley. In: Marić S, Lončarić Z, Florijančić T, Lužaić R, editors. Proceedings of the 45th Croatian and 5th International Symposium on Agriculture, 15-19 February 2010; Opatija, Croatia; 2010. p. 925-929 Available from: file:///C:/Users/Ing%20Strig%C3%A1%C4%8D%20J.PhD/Downloads/Olivera_Stajkovic_Srbinovic_%20tema_BF.pdf

[10] Masayoshi K, Minoru M. Mortar composition. Japanese Patent JP2006315883 A, Publication date 24 November, 2006, Filing date 11 May, 2005, Assignee: Denki Kagaku Kogyo KK, Application number JP2005000137952, [Internet]. Available from: http://www.sumobrain.com/patents/jp/ Mortar-composition/JP2006315883.html [Accessed: 2006-11-24]

[11] Maghchiche A, Naseri R, Haouam A. Uses of blast furnace slag as complex

fertilizer. Journal of Chemistry & Chemical Engineering. 2012;**6**:853-859

[12] Huang Y, Guoping X, Huigao C, Junshi W, Yinfeng W, Hui C. An overview of utilization of steel slag. Procedia Environmental Sciences. 2012;**16**:791-801. (https://doi.org/10.1016/j.proenv.2012.10.108) Available from: https://core.ac.uk/download/pdf/82356046.pdf

[13] Euroslag Technical Leaflet No. 3. Fertilisers from blast furnace and steel slags. Secretary: FEhS- Building Materials Institute. Duisburg, Germany, [Internet]. 2006. Available from: http://www.euroslag.org/fileadmin/_media/images/Research/FACT_SHEETS/Fertiliser_Leaflet.pdf [Accessed: 2006-01-01]

[14] Strigáč J, Števulová N, Mikušinec J, Varečka Ľ, Hudecová D. Antimicrobial efficiency of metallurgical slags for application in building materials and products. Buildings. 2018;**8**:33. DOI: 10.3390/buildings8020033

[15] Grabow WOK, Middendorf IG, Basson NC. Role of lime treatment in the removal of bacteria, enteric viruses and coliphages in a wastewater reclamation plant. Applied and Environmental Microbiology. 1978;**35**:663-669

[16] Remy M. Scientific evidence on the effectiveness of lime. In: Proceedings of the Conference Animal Health and Welfare. A Sustainable Solution from Industry; 18 February 2009; Brussels. Belgium: EuLA c/o IMA-Europe; 2009. pp. 11-21

Section 3

Comprehensive Utilization of Steel Slag

Chapter 3

Treatments and Recycling of Metallurgical Slags

Elena Brandaleze, Edgardo Benavidez and Leandro Santini

Abstract

Steelmaking plants continuously strive to reduce the environmental load in the steelmaking process, resulting in the recycling of energy, water, and other byproducts. In this chapter, techniques for the treatment and recycling of metallurgical slags are described. Metallurgical slags are considered secondary raw materials and are used or added during the process to improve steelmaking practice. Steelmaking slag added into ladle slags makes it possible to minimize slag line wear. BOF-converter slags are also applied in buildup, foaming, or slag splashing practices carried out to prolong the lifespan of refractory lining. Also, EAF slags are commonly used to avoid refractory wear and decrease energy consumption. It is known that cement concrete is one of the most common building materials. Blast furnace crystallized slags are used in cement production, in different percentages. In this sense, understanding the properties of slags is a prerequisite to apply them in different functions. This chapter deals with the measurement and modeling of thermochemical properties of slags, thermophysical properties, and interproperty correlations. Different experimental tests applied in slag characterization are also detailed.

Keywords: slag, recycling, steelmaking, refractory wear, slag properties

1. Introduction

In modern steel plants, the emphasis is placed simultaneously on quantity and quality. Energy consumption per ton of steel production is often considered a benchmark of operational efficiency. Steel production necessarily involves production of millions of tons of slag as well as waste gases containing harmful constituents such as carbon dioxide, dioxin, and furans. Safe disposal of slag and elimination of atmospheric pollution by waste gases are now a matter of serious concern in all steel producing countries. Developed countries are now laying tremendous emphasis on: (a) reduction in energy consumption per ton of steel produced and (b) sustainable environmental management. The most recent thrust is on more integrated green manufacturing and more intensive waste recycling for sustainable development. Efforts are on for economic use of all types of steel plant wastes, such as slag, dust, and flue gases [1]. Slag in ironmaking and steelmaking processes has several metallurgical functions such as preventing contamination by atmosphere, providing thermal insulation, and removing impurities in the liquid metal. Slag as a mixture of oxide has specific properties of melting behavior, viscosity, and surface tension, among others [2]. Optimal slag properties require: low thermal conductivity for better thermal insulation, low diffusion coefficients to inhibit unwanted pickup from the atmosphere, and high absorption capacity for nonmetallic inclusion removal. Physicochemical properties of slag and molten metal directly affect the surface and interfacial properties in refining process of molten metal [2].

Slag is of great relevance in metallurgy processes for steel quality; however, during long years, it was considered an important waste together with the accumulation of refractories after use. This forces us to look for alternative uses to minimize the impact on the environment. A new concept considers to use the waste of one industry as a resource in another industry. One of the main applications that can be given to the slag is as recycled aggregate in the formulation of concrete or abrasive material applications. Nowadays, it is possible to consider the slag as a possible raw material for ceramics, road buildup, or cement production. The mentioned uses include an exhaustive control of the content of Ba, Cd, Cr, Mo, Ni, Pb, V, and Zn, in order to avoid contamination or air pollution. The content of sulfates or sulfides is also controlled to avoid the SO_2 emissions. To avoid volumetric instability of the cement material, MgO content is determined.

The addition of blast furnace slag into cement avoids a part of limestone and coal extraction and makes it possible to diminish the CO_2 emissions that cause environment pollution. It is important to have in mind that blast furnace slags present particular hydraulic properties. They are produced at high temperatures, and under fast-cooling rates, the structure obtained is amorphous. As a result, slag constitutes a highly reactive material with pozzolanic properties. On the contrary, at low-cooling rate, a crystalline structure with a considerable hardness is produced. In this case, slag constitutes a good raw material for cement production and also contributes with Fe addition to the clinker. A great part of the steel produced in the world is obtained through electric arc furnace (EAF), the process involves the use of scrap metal and reduces the CO_2 emissions. EAF slags present great possibilities of recycling, similar to those of BOF slags.

Slags recycling in cement production, road buildup, or other applications (that involve soil contact) are subjected to important environmental requirements. The material must not present reactions or composition changes greater than 1% in a period of 100 years.

One of the biggest challenges for metallurgists is to recycle the slag directly in the steelmaking processes. The operation temperatures and aggressive conditions promote a rapid wear of the refractories of BOF converters and EAF furnaces. Modern LD converters are lined with MgO-C refractories. The presence of carbon controls slag penetration and chemical attack. The capacity of graphite to reduce wear is based upon its large wetting angle for oxide melts. Slags can penetrate the bricks only when the graphite is burnt away near the hot face owing to diffusion of oxygen in between flows during a campaign [1].

In BOF converters, different operation practices using slags as protection media are used to prolong the lifespan of lining refractory:

- Slag coating or buildup, which protects the bottom part of the converter

- Slag foaming, which protects the refractories below the trunnion line

- Slag splashing, which protects the bottom, the middle, and the upper parts of the refractory lining.

In the three cases, the slags require good physical properties (viscosity and surface tension) to achieve good protection results and an adequate slag adherence to MgO-C bricks. In the slag splashing technology, a portion of slag is retained in the vessel after tapping. Slag with low FeO and high MgO is desirable. The practice is carried out with nitrogen injection through a lance at different heights. In this case, some plants require special equipment installation.

Slag foaming is also applied in EAF to prevent the heat loss and to decrease the refractory wear. To improve the performance of these industrial practices and their applications, it is necessary to deeply understand the fundamental phenomena so as to determine slag's characteristics and physical properties required at process conditions.

Another type of slag that is produced in considerable quantity (ton/year) by the steelmaking industry is ladle slag. Because ladle slag is a premelted flux, when it is recycled in the steelmaking process, it is easy to be remelted. This type of slag has good physical properties, high inclusion absorptivity, and high sulfide capacity. Choi et al. [3] described two possible methods to recycle ladle slag: (a) by pouring molten ladle slag to another ladle, at time (taking advantage of the heat) and (b) by making a ladle slag ingot. The latter option mentioned is complex because of the unstable method used for making ingots. Ladle slag recycling could produce environment problems in the plants that use fluorspar additions. HF emissions generated increased F concentrations in the plant water and exhaust gas. However, it is relevant to highlight that ladle slag reuse promotes dephosphorization reaction at BOF and can substitute expensive steelmaking agents, and besides, the process can be more ecofriendly.

To summarize, slag recycling requires the availability of detailed and precise information on their behavior, physical properties, and structural characteristics. This chapter is focused on the chemical composition, structural characteristics, thermal and physical properties, and behavior (at high temperature) of different slags used to protect refractories in BOF and EAF as recycling possibilities.

2. BOF slags recycling practices

Longer lifespans of BOF lining and greater availability result in a reduction of costs per ton of liquid steel. Different steel plants, depending on their lay out, process conditions, and operation, have developed and implemented various techniques for protecting MgO-C refractory lining: brick patching, gunning, manual splashing, chemical splashing, slag coating, slag foaming, and slag splashing [1, 4]. In this chapter, two main slag practices are described: slag coating and slag foaming considering experimental data and theoretical information.

2.1 BOF slags applied in slag coating practice

The application of MgO saturated coating slags helps to reduce the impact of the BOF operation on the refractory lining. These slags form a protective layer on the bottom of BOF MgO-C bricks improving the lifespan of the lining. Slag fluidity and the physics and chemistry of the refractory-slag interaction are relevant to the design of these coating slags. The blowing practice, the end of blow temperature, the ratio of solid-to-liquid phases in the slag, its morphology, its particle size, and chemistry are determining factors for the required properties. Slag-refractory adherence mechanisms are analyzed, and measurements of slag fluidity at high temperature and other physical properties obtained by thermodynamic simulation (up to 1700°C) are informed. Slags of appropriate characteristics allow lining recovery, reducing corrosion damage, oxidation, and erosion of the refractory, as well as the impact of the charge, without altering the metallurgical function of the BOF. Slags used for this purpose need to be optimized and characterized [4].

Slag behavior at process temperatures depends on slag chemistry, melting behavior, viscosity, surface tension, fluidity, FeO content, and MgO saturation. In addition, solid and liquid phases of the slag (at practice temperatures) in the system have relevant influence on the adherence behavior. All these factors lead to/foster the achievement of a good slag layer on the surface of MgO-C bricks.

Experimental tests of physical properties as viscosity and surface tension of multicomponent systems at high temperatures (higher than 1500°C) are difficult to carry out due to the complexity of the process and the amount of time involved. The presence of impurities and possible reactions between the crucible material and the slag produces a transportation problem in the interface that can reduce the values of the properties and can occasionally modify wettability. These factors induce the development of mathematical models and thermodynamic simulation applications to predict the viscosity of a system as a function of slag chemistry [5–8]. The Urbain theoretical model is being one of those more used for the steel-making slags [5], being the Urbain model one of the most widely used for industrial slags. The model estimates the viscosity of liquid silicates and alumino-silicates in function of chemical composition and describes temperature dependence of viscosity through the Weymann expression, Eq. (1):

$$\eta = AT \exp\left(10^3\ B/T\right) \tag{1}$$

where η is the viscosity in poise; T is the absolute temperature in K; and A and B are the parameters that depend on slag chemical composition (A is expressed in poise K^{-1} and B in K), Eq. (2).

$$-\ln A = 0.2693B + 11.6725 \tag{2}$$

The Urbain model [5] classified the oxides into three groups:

- Glass formers: $X_G = X_{SiO_2} + X_{P_2O_5}$
- Modifiers: $X_M = X_{CaO} + X_{MgO} + X_{Na_2O} + X_{K_2O} + 3X_{CaF_2} + X_{FeO} + X_{MnO} + 2X_{TiO_2} + 2X_{ZrO_2}$
- Anfoters: $X_A = X_{Al_2O_3} + X_{Fe_2O_3} + X_{B_2O_3}$

In this chapter, viscosity values of end-of-blow slags were estimated taking into account the Urbain model and applying the viscosity module for melts of FactSage 7.2 software [6]. It is important to mention that during the last year, large thermodynamic databases of multicomponent oxide systems have been developed, and parameters have been optimized in order to reproduce all reliable experimental data within experimental errors. The information obtained through FactSage 7.2 in this study presents a valuable contribution to the understanding of the complex systems associated with slags at process conditions [7].

The industrial end-of-blow slags selected for this study were divided into five groups with different FeO contents: A (18 < FeO < 22), B (22 < FeO < 25), C (25 < FeO < 30), D (FeO>30) and E (FeO<18). Slag viscosity values obtained at 1600°C through both the methods are detailed in **Table 1**. In **Figure 1**, a comparison of the viscosity values obtained for slags A, B, C, D, and E shows that slags of the E group, with FeO content lower than 18 %wt. and a binary index (BI) 2.88, present the highest viscosities. In **Figure 2**, it is possible to evaluate the impact of slag FeO content on the viscosity evolution.

Slag	SiO_2 (%)	CaO (%)	FeO (%)	MgO (%)	MnO (%)	Al_2O_3 (%)	P_2O_5 (%)	S (%)	η_{FS} (poise)	η_{Urbain} (poise)
Group A (18 < FeO < 22)										
1	15.9	44.3	19.6	10.2	6.3	0.8	2.8	0.06	0.25	0.78
2	15.4	47.0	18.7	8.4	5.9	1.5	2.7	0.08	0.26	0.80
3	14.8	45.3	21.6	8.4	6.3	1.0	2.4	0.06	0.24	0.75
4	15.3	45.7	20.7	7.6	6.2	1.9	2.5	0.08	0.25	0.81
5	15.3	45.1	21.0	7.6	6.4	2.1	2.5	0.07	0.25	0.83
Group B (22 < FeO < 25)										
6	14.2	44.1	22.5	8.8	6.3	1.4	2.4	0.07	0.23	0.75
7	14.7	43.9	22.5	7.9	6.4	1.4	2.7	0.07	0.23	0.77
8	14.0	44.3	24.4	7.5	5.8	1.6	2.5	0.08	0.23	0.75
9	15.3	43.8	23.7	7.2	6.2	1.6	2.5	0.08	0.23	0.80
10	14.4	44.2	24.1	7.2	6.1	0.9	2.6	0.07	0.22	0.74
Group C (25 < FeO < 30)										
11	13.2	44.6	25.6	7.5	5.6	1.2	2.3	0.08	0.21	0.70
12	13.8	43.7	26.3	6.9	5.5	1.2	2.1	0.09	0.22	0.73
13	12.1	44.3	25.8	7.9	5.7	1.6	2.2	0.08	0.21	0.68
14	13.9	44.0	25.3	6.9	6.2	1.2	2.5	0.07	0.22	0.73
15	13.4	43.7	26.0	6.9	5.9	1.6	2.3	0.08	0.21	0.73
Group D (FeO > 30)										
16	12.7	39.7	32.1	7.6	5.8	0.8	2.2	0.08	0.18	0.67
17	10.5	32.1	39.9	8.5	6.0	0.8	1.8	0.07	0.15	0.59
18	10.5	42.3	31.4	6.5	5.6	1.2	1.9	0.09	0.18	0.60
19	12.0	40.0	32.5	6.4	5.6	0.9	2.0	0.08	0.17	0.65
20	11.5	39.9	33.6	6.7	5.3	0.7	1.8	0.07	0.17	0.62
Group E (FeO < 18)										
21	18.5	45.2	16.7	7.8	6.8	1.0	3.1	0.06	0.29	0.90
22	13.0	48.2	16.4	9.8	6.0	1.7	2.8	0.09	0.26	0.72
23	18.2	49.9	14.1	6.8	5.6	1.6	3.3	0.09	0.31	0.91
24	17.9	49.0	15.3	6.8	5.9	1.8	3.0	0.10	0.30	0.92
25	17.7	49.1	16.1	6.4	5.8	1.7	2.9	0.10	0.30	0.90

Table 1.
End of blow slag chemistry.

A slag sample with FeO ≈ 18, MgO ≈ 11, CaO ≈ 43, Al_2O_3 ≈ 1.6, and SiO_2 ≈ 17% wt. was characterized by different experimental tests. The melting behavior was determined by hot-stage microscopy (HSM). The critical temperatures obtained are $T_{initial}$ = 1398°C, $T_{softening}$ = 1408°C, $T_{hemisphere}$ = 1412°C, and $T_{fluidity}$ = 1419°C. At 1600°C, the average of the contact angle measured was $\theta_{average}$ = 33°. Also, a good adherence behavior on MgO-C samples was obtained by means of dipping tests. Viscosity values estimated by the Urbain model and FactSage 7.2 at 1600°C for the slag were η_{FSage} = 0.29 poise/η_{Urbain} = 0.90 poise. The solid and liquid phases predicted (at 1600°C) by the Equilib module of the software are 92.25 g of a liquid

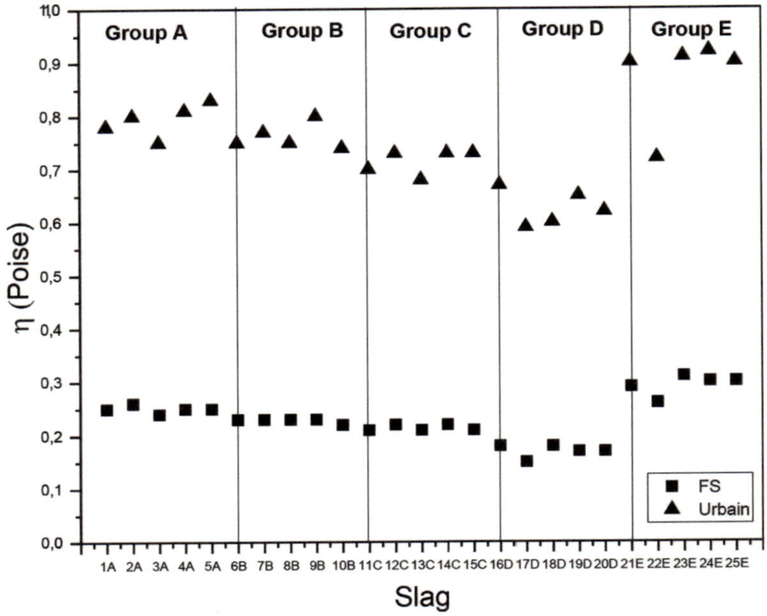

Figure 1.
Comparison of viscosity values of slags for groups A, B, C, D, and E with different FeO content.

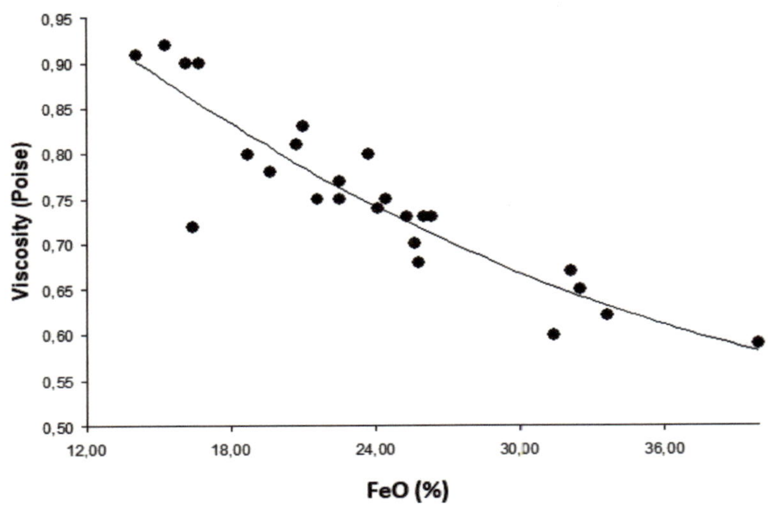

Figure 2.
Influence of iron oxide content on viscosity at 1600°C.

phase L_1 with a chemical composition: 4.70% MgO, 17.4% FeO, 6.47% MnO, 18.42% SiO_2, 44.87% CaO, 2.16% Al_2O_3, 0.98% Fe_2O_3, 3.21% $Ca_3(PO_4)_2$, 0.40% $Mg_3(PO_4)_2$, and 1.12% $Fe_3(PO_4)_2$ and two solid phases: 6.4 g of MgO and 0.33 g of Fe (bcc).

For slags with FeO ≈ 25.8%wt, MgO ≈ 7%wt., CaO ≈ 44%wt., Al_2O_3 ≈ 1%wt., and SiO_2 ≈ 14%wt., the liquid phase L_1 increases up to 97 g, and the chemical composition is as follows: 4.33% MgO, 24.2% FeO, 6.16% MnO, 14.45% SiO_2, 43.8% CaO, 1.03% Al_2O_3, 1.23% Fe_2O_3, 2.82% $Ca_3(PO_4)_2$, 0.33% $Mg_3(PO_4)_2$, and 1.40% $Fe_3(PO_4)_2$. The solid phases include 2.66 g of MgO and 0.44 g of Fe (bcc).

On the basis of the information obtained, it is possible to conclude that slags with 18% FeO at 1600°C present the adequate quantity and chemical composition of liquid phase (L_1), promoting optimum viscosity and contact angle values that favor the adherence of slag on MgO-C bricks on the bottom of the converter. However, slags with higher FeO content promote the increase of the liquid phase proportion (L_1) with chemical composition changes that produce disadvantages regarding wettability. Slags with the chemical and physical properties mentioned are applied in BOF slag coating practice in the industry with good results increasing the lifespan of the lining and protecting the bottom of the converter.

2.2 BOF slags used in foaming practice

Slag foaming constitutes another alternative to use the converter slags to protect MgO-C bricks below the trunnion line [8, 10]. It is important to take into account that in order to enhance the performance of this industrial practice, it is necessary to understand deeply the associated fundamental phenomena. Along the steel conversion process, the oxygen blown is mostly combined with elements dissolved in the melt. The reaction products are oxides, which are finally incorporated into the slag. If a carbonaceous material is added after blow, C reacts with FeO and generates CO, Eq. (3), causing slag foaming.

$$C(\text{injected}) + (FeO) \rightarrow Fe + CO \tag{3}$$

CO for foaming is also produced by decarburization of the metal given by Eq. (4):

$$C(\text{in metal}) + \tfrac{1}{2}\, O_2 \rightarrow CO \tag{4}$$

Gas bubbles in the slag foam could be small or large. The size of the bubbles determines the foam behavior. Foam with small bubbles is dense avoiding the risk of slopping and promoting good slag adherence to the refractory surface. On the contrary, large bubbles like soap generate unstable foam and the likelihood of slopping. Two types of foams can be distinguished: (i) foams initially formed by spherical bubbles that are separated by thick films of liquid and (ii) polyhedral foams in which the bubbles are deformed because of drainage of the liquid film [9–12]. Drainage phenomena decrease the thickness of the liquid separating film to values lower than 1 μm.

As it is known, converter slags are an extremely complex mix of oxides that determine the slag physical properties such as viscosity, density, surface tension, and others. Therefore, accurate viscosity values are essential for the optimization and improvement of metallurgical processes. Experimental measurements have difficulties associated with the high temperature needed (or the high melting point of some slags). Consequently, an alternative is the application of theoretical models to estimate the viscosity and surface tension, as a function of temperature, on the basis of slag chemical compositions [13, 14].

As it was already mentioned, viscosity is considerably affected by FeO content. If the FeO increases in the slag, the viscosity decreases and causes foam bubbles to drain more rapidly, resulting in foam decay. Hence, there is a critical FeO content below which foaming increases and above which the foam is less stable. Surface tension consists in a physical property that depends principally on the surface and not on the bulk. This is affected by the concentration of surfactants and the chemical activity of the surface active component, which determines the surface active concentration in the surface layer [15]. Both properties (viscosity and surface tension) are necessary to estimate the foaming index Σ and to increase the knowledge about this practice. Different expressions have been proposed for foaming index Σ. In this study, Eq. (5) detailed by Fruehan and Matsura [16], is considered:

$$\Sigma = 115\frac{\mu^{1/2}}{\sigma^{0.2}\rho D_B^{0.9}} \tag{5}$$

where D_B, μ, ρ, and σ are the foam bubble diameter, slag viscosity, slag density, and slag surface tension, respectively.

Another point of interest is the foam life time, because it can vary greatly depending on the surfactant compounds present. Transient foams last for a few seconds, while stable foams have prolonged shelf life [12]. In this chapter, results of 16 BOF slags characterized for foaming practice are informed. Slag viscosity and surface tension were determined by theoretical models in order to establish the foaming index applying Eq. (5). The Urbain model [5] and Zaharia, Sahajwalla and Khanna model [17] were used to determine viscosity and surface tension, respectively. The results were completed with information about the thermal behavior of the slags obtained by hot-stage microscopy (HSM). Also, based on the slag chemical compositions, the isothermal solubility diagram (ISD) was applied in order to predict a good foaming behavior and capability of refractory protection.

Another contribution of this work is the information concerning different carbonaceous materials used to foam BOF slags. Three materials were studied considering physicochemical properties and structural aspects. The results were correlated with the information of the slag characterization.

Table 2 details the chemical composition of the BOF slags selected for the study in order to improve the foaming process. Based on the slag chemical composition, the binary basicity index (IB2) and ternary basicity index (IB3) are calculated to predict foamability of the slags applying ISD (see **Table 3**). The values of IB2

Slag	% SiO_2	%CaO	% Al_2O_3	%MgO	%FeO	%MnO	%P_2O_5	%S
1	13.96	42.60	3.04	12.39	22.18	5.96	1.70	0.11
2	13.55	42.97	2.89	8.58	23.56	5.31	1.79	0.09
3	12.31	42.45	2.60	8.28	25.79	7.76	1.76	0.10
4	13.12	46.10	2.26	8.95	21.46	7.06	1.86	0.09
5	16.20	42.41	1.63	10.83	20.71	6.64	1.87	0.07
6	18.09	42.02	2.79	10.08	19.99	6.79	1.71	0.08
7	12.74	41.13	2.25	10.96	26.40	5.79	1.61	0.06
8	17.23	44.32	3.13	9.43	19.15	5.47	2.08	0.09
9	16.74	44.02	2.34	8.66	20.47	7.73	2.04	0.11
10	13.84	40.95	2.14	7.99	25.65	8.33	1.81	0.09
11	12.21	38.19	2.62	7.20	29.55	7.68	1.77	0.09
12	14.47	40.16	1.99	8.26	28.34	5.79	1.60	0.07
13	13.03	39.23	2.67	8.32	26.22	9.96	1.66	0.10
14	14.91	42.47	2.46	9.52	24.05	5.41	1.78	0.09
15	19.68	44.61	3.86	9.47	17.20	5.28	1.96	0.09
16	17.15	45.13	2.57	8.07	19.95	6.67	1.92	0.09

Table 2.
Chemical composition of foaming slags.

Slag	IB2	IB3	η (1550°C) (Poise)	γ (1550°C) (mN/m)	ΣD_b = 0.015 m	ΣD_b = 0.005 m
1	3.05	2.51	1.14	665	4.97	14.97
2	3.17	2.62	1.17	648	5.17	15.57
3	3.45	2.85	1.04	646	4.88	14.72
4	3.51	3.00	1.06	642	4.96	14.96
5	2.62	2.38	1.20	638	5.32	16.02
6	2.32	2.01	1.41	630	5.84	17.58
7	3.23	2.74	1.03	625	5.03	15.15
8	2.57	2.18	1.40	630	5.82	17.52
9	2.63	2.31	1.29	621	5.66	17.06
10	2.96	2.56	1.12	625	5.24	15.79
11	3.13	2.57	1.07	623	5.14	15.49
12	2.78	2.44	1.14	619	5.34	16.09
13	3.01	2.50	1.10	627	5.18	15.59
14	2.85	2.45	1.19	616	5.48	16.52
15	2.27	1.90	1.61	617	6.37	19.18
16	2.63	2.29	1.34	618	5.80	17.47

Table 3.
Basicity index, viscosity, surface tension and foaming index values.

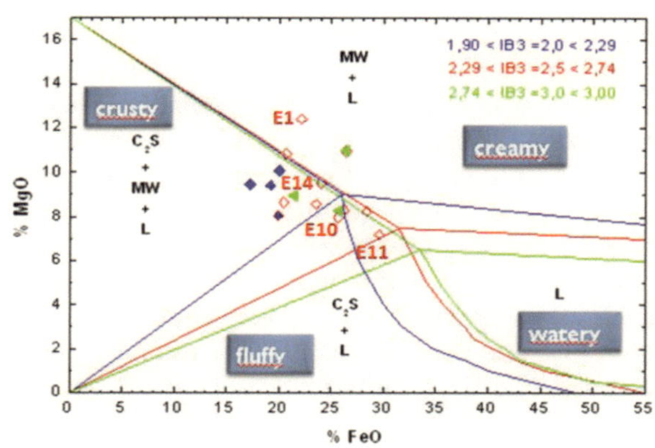

Figure 3.
ISD (%MgO-%FeO) for three ternary basicity index ranges.

indicate that the slags are saturated in MgO and are compatible with refractory lining. All the slags were classified into three ternary basicity index (IB3) ranges: 1.9 > IB3 > 2.29, 2.29 > IB3 > 2.74, and 2.74 > IB3 > 3 and then placed in ISD (%MgO-%FeO).

As it is known, a creamy slag is the optimum condition to obtain the good foamability described by Pretorious [18]. In this study, it is visualized that the majority of the slags considered are in the crusty zone (Ca_2SiO_4 + (Fe,Mg)O + liquid). Only four slags (E1, E10, E11, and E14) with IB3 ~ 2.5 are near the creamy zone ((Fe,Mg)O + liquid; see **Figure 3**). Nevertheless, when ISD includes the % MnO, the same four slags (E1, E10, E11 and E14) are moved toward the right

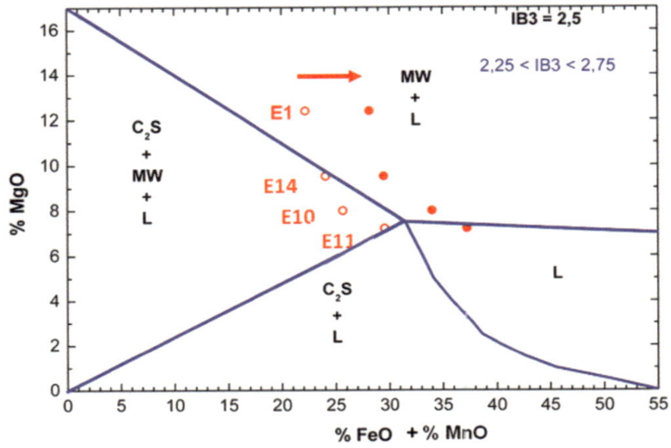

Figure 4.
ISD (%MgO-FeO + %MnO) for IB3 = 2.5.

showing the MnO effect (**Figure 4**). MgO content in the slags mentioned is between 7.2 and 12.4%.

Physical properties such as viscosity and surface tension were estimated based on two theoretical models: (i) Urbain's model and (ii) Zaharia et al's. model, respectively, in order to predict the foaming index (Σ) at 1550°C (see **Table 2**). Two different diameters (D_B) of bubbles were considered in the Σ calculus: 0.015 m and 0.005 m. The results obtained are also detailed in **Table 3**. For slags (E1, E10, E11, and E14) with good foamability properties, the viscosity (η) values are between 1.07 and 1.19 poise and the surface tension between 616 and 665 mN/m, at 1550°C. The results are in agreement with Pretorious [18], who suggests low or medium values of surface tension in the slag for good foaming behavior. At 1550°C, the predicted average foaming index values for two bubble diameters (D_B = 0.015 m and D_B = 0.005 m) are 5.2 and 15.7, respectively. The content of %FeO in these four slags is between 22 and 30% and of %MnO between 5.4 and 8.3%. This information represents a good starting point to optimize the BOF foaming industrial practice. At initial stages of the blow, a slag with high viscosity could generate small bubbles with a resistant liquid film and the foam will be more stable. The correlation between IB3 and Σ shows that the foaming index decreases (slightly) as the basicity index increases (**Figure 5**).

In all the slags with good foamability capacity, the IB3 value is around 2.5. It was observed that when the bubbles present a D_B = 0.005 mm (small size), MnO content (for MnO < 8%) results in random foaming index values compared to the effect of FeO content. The highest foaming index Σ was obtained for slag E15 with 19.68% SiO_2, 17.20% FeO, 9.47% MgO, and 5.28% MnO. FeO content in the slag constitutes a variable that enables to control slag foaming. The study shows the sensibility of viscosity and surface tension, regarding FeO and MnO contents. Also, the higher the viscosity the higher the foaming index.

Slag samples, obtained pre (B_f) and post (A_f) foaming practice, were characterized. The melting behavior was determined by hot-stage microscopy (HSM). **Figure 6** shows the average critical temperatures: softening temperature (T_s), hemisphere temperature (T_h), and fluidity temperature (T_f) of the samples.

It is worth noting that critical temperatures in slag samples increase after the foaming practice. These results indicate a dynamic evolution of the slag chemical composition during the practice. Average critical temperatures range for B_f (before

DOI: http://dx.doi.org/10.5772/intechopen.80595

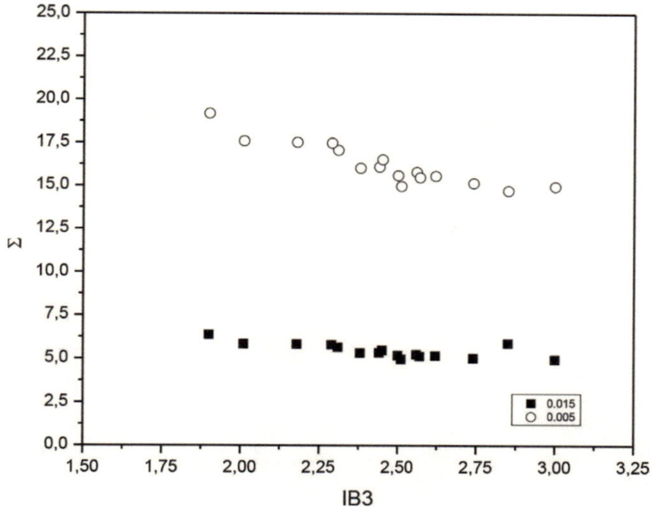

Figure 5.
Incidence of IB3 on foaming index Σ.

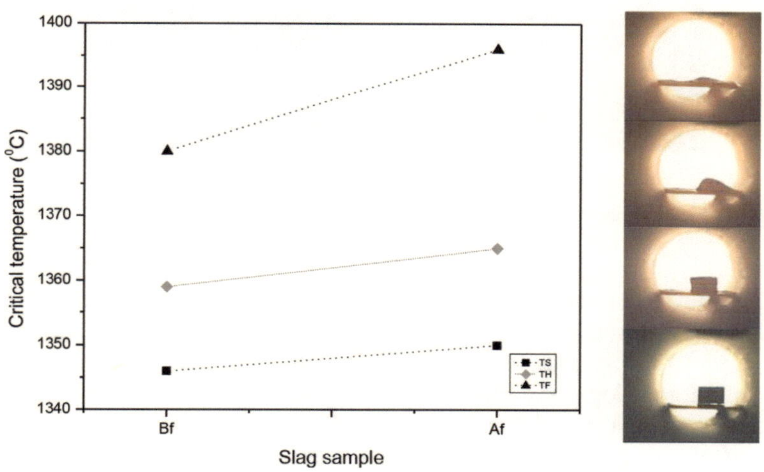

Figure 6.
Average critical temperature values (melting behavior) of the slags before and after foaming practice.

foaming) slags are between 1345 and 1380°C. However, for A_f slags (after foaming), an increase ≈ 3% of critical temperatures was determined.

During slag foaming, when carbon is injected into the slags, gas bubbles are formed around the carbonaceous particles. It is necessary to take into consideration that different carbonaceous materials could produce different impact on slag foaming. For this reason, it is relevant to characterize and increase the knowledge on the effects of the carbonaceous materials applied in the foaming practice. Chemical and structural characteristics of three carbonaceous materials were studied. In **Table 4**, results of the chemical characteristics of carbonaceous materials are informed. Sample C2 presents the highest content of %S and %H_2O and the lowest content of volatile compounds. Ash content is around 11% in materials C1 and C2 (typically in metallurgical cokes). Nevertheless, material C3 contains very low percentage of ash because it is a petroleum coke. Moreover, it presents the highest carbon content. Sulfur content is similar in all samples.

Carbonaceous material	% H_2O (at 105°C)	%Ash	%Volatile	%C	%S
C1	0.76	11.81	1.84	86.35	0.62
C2	1.02	10.95	1.38	87.67	0.70
C3	0.61	0.20	2.24	97.56	0.63

Table 4.
Chemical characteristics of carbonaceous materials.

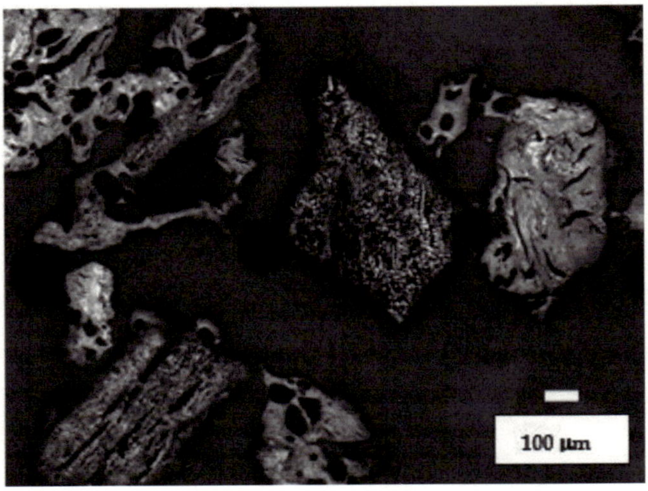

Figure 7.
Different types of porous carbonaceous particles identified.

Size distribution of carbonaceous particles was determined by ASTM sieves. It was observed that sample C1 has particles of small sizes, sample C2 also contains particles of up to ≈ 0.75 mm, and sample C3 presents particles between 0.80 and 2.5 mm. The morphology, the phases present, and the porosity of the carbonaceous particles were studied by light and scanning electron microscopy including EDS analysis. Different types of carbonaceous particles were identified in all the samples (**Figure** 7). Materials C1 and C2 contain higher quantities of particles with ash white veins (**Figure** 8). This result is consistent with the %Ash experimentally determined.

The morphology, size, and distribution of pores in each type of carbonaceous particles are variable. There are irregular, spherical, and elongated pores of different sizes and proportion in the particles observed. It was possible to establish that sample C3 has the largest proportion of pores with different sizes (800–50 μm), and samples C2 and C1 present porous with sizes between 350 and 50 μm. Rahman et al. [19] informed that metallurgical cokes (such as materials C1 and C2) promote gas emissions one order higher (in magnitude) than the synthetic coke. This is explained by the higher chemical interaction in contact with the slag, due to the reduction of the oxides present in the ash and FeO content. To control foam height, carbonaceous particles can be added to slag with adequate size and low volatile content. This is another alternative to avoid the risk of slopping. It is important to highlight that the foaming index decreases as the size of carbonaceous particle increases. Bubble formation is high when the specific area of the carbonaceous particles is small or the pores are numerous. On this base, material C3 is the most favorable for the slag foaming practice due to the size and distribution of the pores.

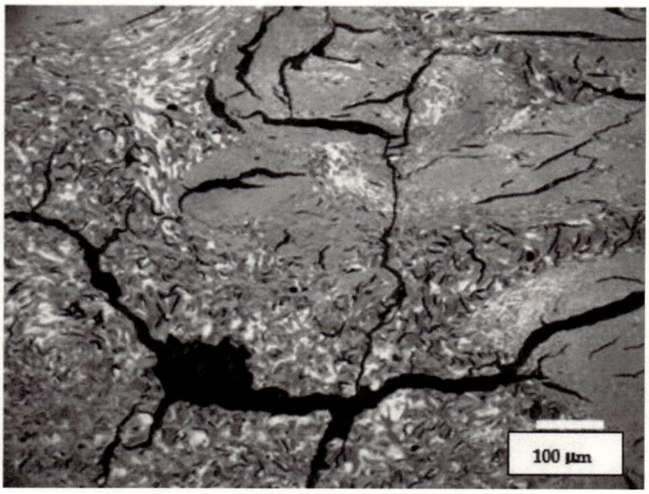

Figure 8.
Detail of one particle with considerable white ash veins.

Furthermore, the ash content of materials C1 and C2 allows predicting rapid gas generation, which is not favorable for foam stability.

Based on these results, it is possible to confirm that the evolution of properties (viscosity and surface tension) of BOF slags during the operation represents the key to obtain the best aptitude for foaming practice, avoiding the risk of slopping. High viscosity is perhaps the most obvious factor in the stabilization of BOF-converter foams, and it retards the rate of drainage in the films of the bubbles. Nevertheless, surface tension is also a relevant physical property that should be taken into account to predict foam behavior. Both properties, high viscosity and surface tension, are determined by the chemical composition and the evolution of the slag at process conditions. Carbonaceous materials also exert an effect in the foaming practice; it is possible to recommend the addition of petroleum coke (material C3) in the foaming practice, because of its ash low content, high proportion of large pores, and higher carbon content.

3. EAF slags reused in the foaming practice

Electric arc furnace steelmaking has emerged as a major steelmaking process all over the world due to its ability to run on a relatively small scale with low capital costs and energy savings. Steel scrap is the preferred metallic charge material, and sponge iron is also used regularly in most plants in order to dilute tramp elements introduced through the scrap [1, 17]. The growth of the arc furnace-based steel industry has been encouraged by modern arc furnaces of large capacity and ultra-high power, electrical efficiency improvement, and metallurgical efficiency through oxygen lancing or the co-jet technology. In conventional EAF steelmaking, natural gas is generally used as a supplementary energy source. Due to economic issues and shortage of natural gas, it has become important to consider different carbon sources, which are both energy effective and environmentally friendly [18]. In addition, the decrease of refractory consumption is a crucial topic of consideration for all steel plants.

Not only does the slag foaming practice protect the equipment covering the arcs and the refractory lining (the furnace roof and sidewalls from excessive heating and radiation), but it also reduces the arc noise, decreases electrode and electricity consumption, and has present a significant impact on improving thermal efficiency. The foamy slag provides an insulating layer to the melt, thereby reducing energy loss. Slag foaming involves the expansion of molten slag by CO gas bubbles evolving from chemical reactions at the slag-metal interface. Slag viscosity and surface tension control the movement of bubbles in the liquid. Foaming was generally found to improve as surface tension decreases, increasing slag viscosity and suspension of second-phase particles. The reaction between FeO present in the slag and carbon is strongly endothermic. It is important to note that sulfur additions suppressed slag foaming and tended to increase the size of CO gas bubbles. An increase in silica concentration promotes lower surface tension of slag and leads to smaller CO gas bubbles and their easier escape from the slag. The injection of oxygen and carbonaceous materials in industrial EAF furnaces creates highly dynamic and nonsteady state conditions. Oxidation and reduction reactions continuously change slag composition, and chemical reactions produce CO gas that also changes significantly with time [19].

In agreement with BOF foaming practice, the rate of gas generation was found to be an important parameter in carbon-slag interactions and needs to be optimized to maintain optimum levels/duration of gas entrapment by the slag and foaming behavior. The type of carbonaceous material added affects the foaming practice. Metallurgical cokes affect foaming because of the ash content. Natural graphite, on the other hand, produces excellent slag volumes but slow reduction of iron oxide. Slower rates of gas generation and higher surface tension values cause slag to trap gases and sustain foaming. Slag foaming height is well connected to carbon injection rate and slag chemical properties. Bubble sizes in EAF slags are ≈ 1–2 mm. In general, bubble number in the slag foam is greater than the number of coal particles. One particle of coal can form more than one bubble. In EAF steelmaking, the foam becomes less stable toward the end of the process.

4. Reused of ladle slags

The amount of ladle slag generated in the steelmaking is important. The former method of recycling ladle slag is pouring the molten slag from a ladle to another or making ladle slag ingot [3]. The ingot production is complex. However, nowadays three main types of ladle slag recycle products are used depending on the particle size and chemical composition: (a) desulfurizing agent in hot metal pretreatment, (b) as a product to substitute converter flux, and (c) as a product to substitute ladle flux. It is possible to reduce steelmaking cost and get ecofriendly benefits, recycling ladle slag.

Ladle slags present similar melting behavior respect synthetic flux ($T_s \cong 1334°C$, $T_f \cong 1346°C$) and also have a good inclusion absorptivity. For this reason, they constitute an alternative for ladle flux substitution. Slag low-melting point possibilitates the application in ladles during secondary refining process reducing the fluorspar or lime additions. Finally, lumps of slags could be added in BOF to improve dephosphorization.

5. Slags reused as concrete raw material

Slag is a partially vitreous byproduct of the process of smelting ore, which separates the desired metal fraction from the unwanted fraction. Blast furnace slag is furthermore classified into granulated blast furnace slag and air-cooled blast furnace slag. Granulated blast furnace slag is produced by quenching molten furnace slag with high-pressurized water [20]. Blast furnace slags cooled in air constitute a crystallized material used as raw material instead of sand in the production of concrete. In the case of water applied during granulated slag production, it is possible to recover heat from the slag quenching water, thereby reducing energy consumption. Blast furnace slag is obtained at 1300–1400°C; viscosity is around 0.4–1.0 poise; the chemical composition mainly includes FeO, CaO, MgO, SiO_2, Al_2O_3, TiO_2, S, and C; the alkalinity (CaO/SiO_2) is 1.2; and the alkalinity [$(CaO + MgO)/SiO_2$] is around 1.5. A lot of research has focused on the reuse of blast furnace slag to make building materials. Crystallized slag sand is added to Portland cement in particles of ≈ 3.15 mm. The chemical and mineralogical composition of Portland cement is detailed in **Table 5**. Senani et al. [21] informed that concrete with crystallized slag sand addition presents good physical and mechanical properties. This study confirms that the additions of blast furnace slags (0–4 mm of size) in concrete (up to 20–25%) result in good compressive strength (≈30 MPa after 28 days) for the grades of concretes tested.

The use of blast furnace slag as artificial rocks is also possible because they are similar to semi-hard stones. As a recyclable material, they support resource conservation by substituting nonrenewable natural stone. Land and marine applications include seaport and airport civil engineering projects, coastal protection structures, seaweed bed rehabilitation, and land surface coverage [22]. This type of slag may represent a good alternative to reduce the cost of cement production, contribute to environment protection, and provide enormous social and economic benefits.

Brand and Roesler studied different basic steel slags: energy optimizing furnace (EOF), electric arc furnace (EAF), and arc oxygen decarbonization (AOD), as aggregates in concretes [23]. However, for civil applications, exhaustive controls are carried out in order to determine contents of Ba, Cd, Cr, Mo, Ni, Pb, V, and Zn, in order to avoid soil contamination. The content of sulfates or sulfides is also controlled to avoid SO_2 emissions causing air pollution. It was found that steel slag is volumetrically unstable as compared to blast furnace slag due to the content of expansive oxides such as MgO and CaO. In the case of basic slags, the authors propose to modify the surface of the aggregates by blending them with a slurry mixture of fine quarry dust and cement in order to avoid the expansion and to reduce the porosity. Another possibility is the use of steel slag as a binder or filler replacement in composite materials in civil engineering production. These binder or

Chemical composition of Portland cement	Range (%)	Mineralogical composition	Range (%)
CaO	56–63	C_3S	50–65
Al_2O_3	4–6	C_2S	10–25
SiO_2	19–27	C_3A	9–12
Fe_2O_3	2.5–3.5	C_4AF	7–11
MgO	1–2		
Na_2O	0.1–0.6		
K_2O	0.3–0.6		
Cl^-	0–0.2		
SO_3	2–3		
CaO	0.5–2.5		

Table 5.
Chemical and mineralogical composition of Portland cement.

filler substitutes were mixed into composites, and their compressive strength was tested with good results.

6. Conclusion

The disposal and exploitation of residues from steelmaking plants are still an open problem because of the huge amount and the remarkable variety of waste materials. Slag recycling in the steel job constitutes a relevant way of reusing with important economic impact on refractory costs. The coating and foaming practices are very useful in BOF converters and EAF. The ladle slag recycling also represents an important alternative to reduce steelmaking cost, decreasing the consumption of fluorspar and lime (among others metallurgical applications), and get ecofriendly benefits.

The growing trend in the construction industry is to develop sustainable buildings. The principles underneath this movement bring new requirements with an emphasis on the rational use of material and energy resources by controlled minimization of total emissions produced. One of the possible ways of achieving sustainable development in the construction industry is to use easily renewable raw material resources and waste materials instead of limited and finite resources. In this sense, the interest shown by the professional community in aggregates based on secondary raw materials is increasing, and slags constitute a good alternative.

Author details

Elena Brandaleze*, Edgardo Benavidez and Leandro Santini
High Temperature Physicochemistry Group, Metallurgy Department, DEYTEMA Center, Universidad Tecnológica Nacional, Facultad Regional San Nicolás, San Nicolás, Argentina

*Address all correspondence to: ebrandaleze@frsn.utn.edu.ar

References

[1] Chakrabaharti AK. Steel Making. 1st ed. PHI Learning Private Limited; 2014. 230 p. ISBN: 978–81–203-3050-4

[2] Jung E-J, Kim W, Sohn I, Min DJ. A study on interfacial tension behaviour between solid iron-$CaO-SiO_2$-MO system. Journal of Materials Science. 2010;**45**:2023-2029. DOI: 10.1007/s10853-009-3046-1

[3] Choi IS, Song WY, Yoon C, Shin KC. The technology of recycling ladle slag. In: Proceedings of the 4th European Oxygen Steelmaking Conference (EOSC '03); 12–15 May 2003. Graz, Austria: EOSC; 2003. pp. 175-181

[4] Madías J, Brandaleze E, Topolevsky R, Camelli S. Slag-refractory adherence mechanisms. Refractories Applications and News. 2004;**9**:21-24. DOI: 10.1.1.200.2234

[5] Urbain G. Viscosity of silicate melts. Transactions and Journal of the British Ceramic Society. 1981;**80**:139-141. DOI: 10.10.1002/srin.19870151310

[6] Bale CW, Belisle E, Chartrand P, Degterov SA, Eriksson G, Gheribi AE, Hack K, Jung IH, Kang YB, Melançon AD, Pelton AD, Petersen S, Robelin C, Sangster J, Spencer P, Van Ende MA. FactSage thermochemical software and databases—2010–2016. CALPHAD: Computer Coupling of Phase Diagrams and Thermochemistry. 2016;**54**:35-53. DOI: 10.1016/j.calphad.2016.05.002

[7] Brandaleze E, Valentini M, Santini L, Benavidez E. Study on fluoride evaporation from casting powders. Journal of Thermal Analysis and Calorimetry. 2018;**133**(1):271-277. DOI: 10.1007/s10973-018-7227-6

[8] Seetharaman S, Teng L, Hayashi M, Wang L. Understanding the properties of slags. ISIJ International. 2013;**53**(1): 1-8. DOI: 10.2355/isijinternational.53.1

[9] Cicutti C, Valdez M, Pérez T, Donayo R, Petroni J. Analysis of slag foaming during the operation of an industrial converter. Latin American Applied Research. 2001;**32**:237-247. ISSN: 1851-8796

[10] Santini L, Pérez J, Rapetto A, Benavidez ER, Brandaleze E. Evaluation of BOF slag applied in foaming practice. In: AISTech Conference Proceedings (Aistech 2013); 6–9 May 2013; Pittsburgh, Pennsylvania, USA. 2013. pp. 2105-2112. ISSN: 1551-6997

[11] Pretorius EB, Carlisle RC. Foamy slag fundamentals and their practical application to electric furnace steelmaking. In: Proceedings of the 16th Process Technology Conference; 1998 Nov 15-18; New Orleans, LA, USA. 1998. pp. 275-292. DOI: 10.1590/1980-5373-MR-2016-0059

[12] Ismail AN, Nur Farhana MY, Jamaludin SB, Idris MA. Influence of recycled wastes on slag foaming during EAF steelmaking. Materials Science Forum. 2015;**819**:381-386. ISSN:1662-9752. DOI: 10.4028/www.scientific.net/MSF.819.381

[13] Kapilashrami A. Interfacial phenomena in two phase systems: Emulsions and slag foaming [thesis]. Stockholm: Royal Institute of Technology; 2004. ISBN: 91-7283-930-9

[14] Zhang GH, Chou KC, Mills KC. Modelling viscosities of CaO-MgO-Al_2O_3-SiO_2 molten slags. ISIJ International. 2012;**52-3**:633-637. DOI: 10.1016/S1006-706X (16)30099-1

[15] Mills KC. The Estimation of Slag Properties. Short Course presented in the Southern African Pyrometallurgy; 2011

[16] Matsura H, Fruehan RJ. Slag foaming in an electric arc furnace. ISIJ International. 2009;**49-10**:1530-1539

[17] Zaharia M, Sahajwalla V, Khanna R, Koshy P, O'Kane P. Carbon/slag interactions between coke/rubber blends and EAF slag at 1550°C. ISIJ International. 2009;**49-10**:1513-1521. DOI: 10.2355/isijinternational.49.1513

[18] Pretorious EB. Slags and the Relationship with Refractory Life and Steel Production. ABM Brasil: ABM Course. Process Technology Group, LWB Refractories; 2003

[19] Rhaman M, Khanna R, Sahajwalla V, O'Kane P. The influence of ash impurities on interfacial reactions between carbonaceous materials and EAF slag at 1550°C. ISIJ International. 2009;**49-3**:329-336. DOI: 10.2355/isijinternational.49.329

[20] Junak J, Stevulova N. The study of washed recycled concrete aggregates and blast furnace slag utilization in concrete production. Advanced Materials Research. 2015;**1100**(3): 197-201. DOI: 10.4028/www.scientific.net/AMR.1100.197

[21] Senami M, Ferhoune N, Guettala A. Substitution of the natural sand by crystallized slag of blast furnace in the composition of the concrete. Alexandria Engineering Journal. DOI: 10.1016/j.aej.2016.05.006. in press

[22] Gu HF, Yao CG. Environmental conservation efforts and byproducts recycling technologies of JFE. Advanced Materials Research. 2012;**573–574**: 379-382. DOI: 10.4028/www.scientific.net/AMR.573-574.379

[23] Sabapathy YK, Balasubramanian VB, Shankari NS, Kumar AY, Ravichandar D. Experimental investigation of surface modified EOF steel slags as coarse aggregate in concrete. Journal of King Saud University- Engineering Sciences. DOI: 10.1016/j.jksues.2016.07.002

Chapter 4

The Comprehensive Utilization of Steel Slag in Agricultural Soils

Angélica Cristina Fernandes Deus,
Rosemary Marques de Almeida Bertani,
Guilherme Constantino Meirelles,
Anelisa de Aquino Vidal Lacerda Soares,
Lais Lorena Queiroz Moreira, Leonardo Theodoro Büll
and Dirceu Maximino Fernandes

Abstract

The use of metallurgical solid wastes such as steel slag, in agricultural activity, has become very important to contribute to reducing the accumulation of such wastes in the environment and to increase crop production. So, this chapter aims to emphasize the main aspects of the application of slags to soil chemical attributes as elevation of pH and neutralization of Al^{3+} toxic in acid soils and increase nutrient content as phosphorus, calcium, magnesium, some micronutrients, and silicon. In addition, the advance in studies of the utilization of these residues in no-tillage systems in tropical soils will be discussed. Aspects related to monitoring the presence of heavy metals will be addressed.

Keywords: calcium and magnesium silicate, soil fertility, steel industry residue

1. Introduction

The metallurgical industries can produce various residues, and some of these residues can be utilized with success in the agricultural activity, as in the case of the steel slag, which has brought important contributions in agricultural production.

In agriculture, slags can be used as fertilizers and corrective of soil acidity [1]. Slags are calcium and magnesium silicates, which show neutralizing action due to SiO_3^{2-} base [2]. Additionally, steel slags have been used as a low-cost source to supply Si to rice plants [3]. In China, the first steel slag fertilizer program invested by Taiyuan Iron and Steel Group and Harsco Corporation of the USA started building in 2011 [4].

Steel slag is the result of industrial processes in which iron mineral is reduced, generating products as pig iron (iron with a high proportion of carbon) and steel. These processes can occur in different types of furnaces such as basic oxygen furnace or electric arc furnace [5]. Inside the furnaces, oxygen pressure is injected to remove impurities such as gaseous carbon monoxide, silicon, manganese, phosphorus, and some iron as liquid oxides; these impurities combine with lime and dolomitic lime to form the steel slag. Steel can also undergo a greater process of refinement, into a ladle and, at the end of the process, to generate ladle steel slag [5]. The steel slags from these two furnaces are very similar. However, ladle steel slag, resulting from further refining, is quite different from steel slag [5]. For production of every tone of steel, nearly 150 kg of slag is generated [6].

There are several types of steel slag, which present variation in chemical composition, physical composition, and solubility, and these variations occur mainly, due to the different processes and different raw materials used in the metallurgical industries. Examples of slags are steel slag, stainless steel slag, ladle furnace slag, and blast furnace slag; all of these residues have already been tested for agricultural use at some moment, as will be shown in the chapter.

2. Steel slag composition

The main chemical components, which slags contain in their composition and which are important for their use in agriculture, are CaO, MgO, SiO_2, P_2O_5, FeO, and MnO. The amount of these components in each slag varies widely depending on raw materials, type of steel made, furnace conditions, and other aspects [4, 5].

Table 1 shows the chemical composition of different steel slags compared to wollastonite which is a rock and is considered as the standard of silicates [7]. The chemical evaluation of soil acidity correctives for agricultural purposes consists of the following determinations: neutralization power (NP) and calcium and magnesium contents [8]. The different slags have different contents about the components that characterize a material as soil acidity corrective that can provide different effects in the neutralization of the soil acidity.

Considering the Brazilian legislation of soil acidity corrective materials [9], for example, all the slags evaluated in **Table 1** can be considered as an alternative source to limestone, because the values of NP and somatory (%) of CaO and MgO are according to the established values of 60% $ECaCO_3$ and 30%, respectively; this means that the application of such materials allows good performance in the neutralization of soil acidity.

We can also observe different concentrations of Si and P_2O_5 in the different slags. LS showed the highest content of Si, while SS has a great amount of P_2O_5 (**Table 1**). These variations may lead to possible increase in the content of these elements in the soil and increase in the availability to the plants.

Table 2 indicates the micronutrients and heavy metal contents in different slags [7]. It is noted that some materials have a higher amount of micronutrients, for example, the steel slag with a higher presence of iron than the other materials and

Materials	CaO	MgO	NP	RR	ECC	Si	P_2O_5	K_2O
	%		$\%ECaCO_3$		%		$g\ kg^{-1}$	
SSS	37.65	9.55	84	71	60	13.6	3.5	0.3
SS	28.13	6.10	70	71	50	14.2	11	0.3
LS	36.10	5.76	77	80	62	21.6	2.5	0.3
W	30.00	3.00	60	100	60	16.0	1.5	0.1

Adapted from [7].
SSS, stainless steel slag; SS, steel slag; LS, ladle slag; W, wollastonite; NP, neutralization power; RR, reactivity rate, expresses the percentage of corrective material that reacts in 3 months; ECC, effective calcium carbonate.

Table 1.
Chemical and physical characterization of different slags.

Materials	Cu	Fe	Mn	Zn	Cd	Ni	Pb	Cr	Hg
	mg kg^{-1}								
SSS	20	38.000	5.300	50	3.1	53.6	12.6	990	<0.1
SS	30	193.500	21.500	70	14.5	3.0	<0.5	941	<0.1
BFS	20	17.400	51.000	50	1.8	1.4	<0.5	104	<0.1
LS	20	28.600	3.700	50	1.6	1.1	<0.5	126	<0.1
W	10	600	100	40	<0.03	0.3	<0.5	0.5	<0.1

Adapted from [7].
SSS, stainless steel slag; SS, steel slag; LS, ladle slag; BFS, blast furnace slag; W, wollastonite.

Table 2.
Micronutrients and heavy metal contents in different slags.

the blast furnace slag with higher content of manganese; the presence of higher micronutrients could be interesting to provide this element for the plant.

The main restriction for the use of slags is the presence of heavy metals in its composition, which have to be evaluated before their soil application. Heavy metal contents in the slags shown in **Table 1** agree with the values which are considered tolerable for application to the soil [10].

3. Steel slag modifying soil chemical attributes

The results of the researchers with the use of steel slag in agriculture have demonstrated that the proper application of these residues brings benefits on the chemical attributes of the soil, such as increase of pH in acidic soils; increase in the nutrient content of phosphorus, calcium, and magnesium; and increase in the content of the beneficial element silicon [6, 11–14]; these aspects contribute to greater crop production.

For proper application, the initial chemical characterization of the soil to calculate the dose either for use as corrective of soil acidity or fertilizers is important. The homogeneous application of the residue in the area and the adequate incorporation in the soil are also important aspects to be considered.

The application of slag can neutralize part of the soil acidity; this occurs because these materials have in their composition the neutralizing base SiO_3^{2-} [15] that reacts in water and releases hydroxyl (OH^-) ions, according to the equations below [2]:

$$CaSiO_3 + H_2O \rightarrow Ca^{2+} + SiO_3^{2-} \tag{1}$$

$$MgSiO_3 + H_2O \rightarrow Mg^{2+} + SiO_3^{2-} \tag{2}$$

$$SiO_3^{2-} + H_2O\ (\text{soil}) \leftrightarrow HSiO_3^- + OH^-\ \ (Kb_1 = 1.6 \text{ x } 10^{-3}) \tag{3}$$

$$HSiO_3^- + H_2O\ (\text{soil}) \leftrightarrow H_2SiO_3 + OH^-\ \ (Kb_2 = 3.1 \text{ x } 10^{-5}) \tag{4}$$

$$OH^- + H^+\ (\text{Soil Solution}) \rightarrow H_2O \tag{5}$$

$$OH^- + Al^{3+}\ (\text{Soil Solution}) \rightarrow Al(OH)_3 \tag{6}$$

The value of the ionization constant (Kb_1) shows that SiO_3^{2-} is a weak base, that is, the OH^- forming reaction is relatively slow and partial. The hydroxyl (OH^-) produced neutralizes the H^+ of the soil solution and the phytotoxic Al^{3+}, and, consequently, there is an increase in pH and a decrease in the concentration of the potential acidity (H + Al) [16].

Correction of acidity with slag occurs similarly to the use of limestone. In addition to the decrease in acidity levels, Ca and Mg supply also occurs in soil [6, 11, 12], with positive effects for crops such as potatoes, rice, black oats, beans, soybeans, alfalfa, coffee, and sugarcane, among others, making steel slag a source capable of being used instead of limestone [11, 13, 17–21, 23].

The application of steel slag was also studied in acidic tropical soils cultivated under no-tillage system. Tropical soils are characterized by being acid with excess of Al^{3+} that is a toxic element to plants because it limits the root development.

The no-tillage system is growing in these areas; however, one of the requirements for the success of this system is that there is no soil revolving, so the acidity corrective material should be applied to the surface without incorporation. Due to the fact that the limes present low solubility in water and little movement in the soil profile [24], the study of the use of slags (silicates) in no-tillage system has increased due to the greater solubility of this residue [2]. Calcium silicate is 6.78 times more soluble than lime [2], so the use of steel slag in the no-tillage system can be an important alternative source for the process of soil acidity correction, because it may promote greater mobility of SiO_3^{2-} anions in soil profile, ensuring faster soil acidity correction in deeper layers in relation to the lime.

Studies conducted by [19, 22, 25–28] showed positive effects in soil acidity correction of surface application of steel slag compared to lime in no-tillage system, in tropical soil.

After 27 months of application of steel slag at doses 2, 4, and 8 t/ha, without incorporation, under no-tillage system, there was an increase in base saturation and

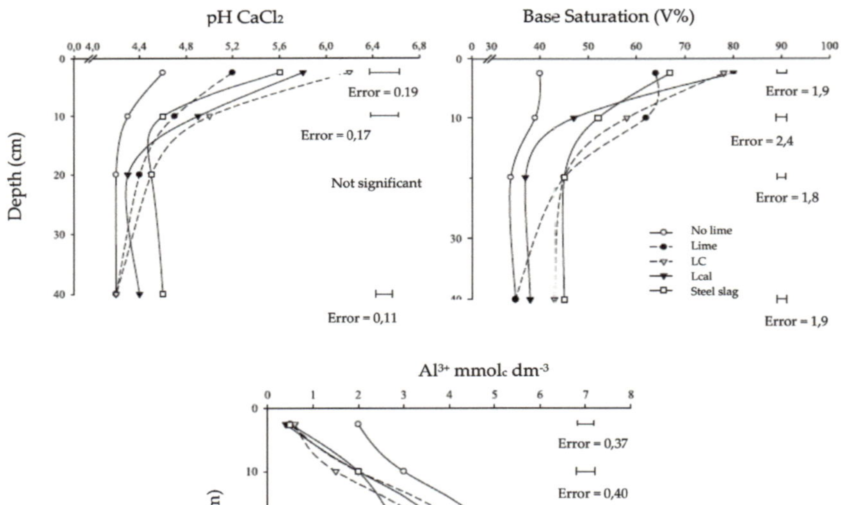

Figure 1.
pH, base saturation (V%), and Al^{3+} content at different depths of a dystrophic Red Latosol at 27 months after surface application of lime, centrifuged sewage sludge (LC), steel slag, and lama cal (Lcal) in the no-tillage system (source: [22]).

reduction of Al^{3+} up to 0.40 m with application of steel slag; in the same period, the effects of limestone on the soil were observed up to 0.20 m (**Figure 1**) [22].

In a long-term experiment under no-tillage system, [26] applied limestone and slag to increase base saturation to 70% and observed that the steel slag corrected acidity and increased the bases up to 0.10 m at 12 months. After 18 months of reaction, the steel slag corrected the soil acidity up to 0.60 m and increased the bases up to 0.40 m, while the effects of the lime were observed until 0.20 m [28].

On the other hand, the study conducted by [7] with surface application of steel slag, blast furnace slag, ladle furnace slag, and stainless steel slag at the installation of the no-tillage system presented that the ladle furnace slag, stainless steel slag, and steel slag had similar efficiency to the lime in the neutralization of soil acidity at 24 months after application. The divergence of the results reported above with the application of slag in no-tillage system can be explained by the chemical composition of these residues, because the industrial process promotes the production of several types of slag, with different recrystallization depending on the amount of Ca and Mg and of the cooling time, aspects that may reduce its solubility [29].

In addition to the corrective effect of the slag, several studies point to the efficiency of slag to increase phosphorus nutrient in the soil [7, 17, 19]. Some researchers attribute the P content increase with the use of slag due to the competition between Si and P that occurs through the same adsorption sites of soil colloids. This occurs due to the application of slag (silicates) to generate a concentration gradient of the anion silicate (SiO_3^{2-}), which removes the P adsorbed to the colloids of Fe and Al oxides from the soil, which under the conditions of the tropical soils are in great quantity [30]. Others relate the P content increase with the increase of the soil pH provided by these materials, as well as greater solubilization of the organic phosphorus and the labile fraction, increasing its content in the form available to the plants. In addition, steel slag has P in its composition, which may contribute to such results, as can be seen in **Table 1**.

The application of steel slag in soil can contribute to the chemical attributes of the soil, reflecting the benefits to the agricultural sector; however, this contribution is variable depending on the composition of the used residue making it always important to know the chemical and physical characterization of the residue before its application to the soil.

The presence of heavy metals in steel slag is one of the problems with its use that at high levels, it can become toxic and limit its use in agricultural activities. The steel residues have very variable chemical composition due to the different processes to obtain them, becoming the important quantification of these elements in the soil and the crops after long-term residue application.

After 18 months of steel slag incubation in an Argisol, it is observed [31] that the solubility of Zn, Mn, Pb, Ni, Cd, and Cu decreased due to the passage of the exchangeable form retained in oxides and residues, explained by the phenomenon of chemosorption influenced by the soil pH. In another study [33], it is found that the application of 8 and 4 Mg ha^{-1} of steel slag, respectively, did not increase the availability of heavy metals of the soil. The increase of heavy metals, such as cadmium (Cd), chromium (Cr), nickel (Ni), mercury (Hg), lead (Pb), and arsenic (As), was insignificant compared to the application of 8 t/ha [32]. The steel slag, ladle slag, and stainless steel slag (chemical characterization in **Tables 1** and **2**) applied as soil acidity correctives, at the dose necessary to raise the initial soil saturation to 70%, did not change the soil metals content evaluated at 12 and 23 months after the application of residues, and the Pb content decreased with the application of residues [7]. Also, the additions of heavy metals were not observed in the plant tissue of beans, soybean, and black oats cultivated in the area where the slags were applied [7, 32]. The presence of elements in the composition of residues in nontoxic amounts and the soil pH increase contributed to avoid toxicity by heavy metals [32].

In the soil, heavy metals can be adsorbed by specific reactions (chemosorption) influenced by soil pH or nonspecific [33]. The pH and CEC and the presence of cations affect the adsorption and ionic speciation of heavy metals in soils [34]. In weathered soils, such as those of the cited studies, the presence of colloids with pH-dependent charges represents more than 70% of the total cation binding sites. Thus, in this soil type, the acidity correction is of great importance in the adsorption of heavy metals [33].

Although the above studies did not observe an increase in the heavy metals' levels in soil and plants, monitoring of the availability of these toxic elements in the soil and in the plant, as well as the characterization of the residue to be used, is fundamental for the success in the residues used in agriculture.

4. Benefits of using steel slag for crops

Slag application favors the increase of pH and the availability of nutrients such as Ca, Mg, and Si in the soil, which leads to the increase in the absorption of these elements by the plant, favoring the growth and yield of the crops. Slags application may supply silicon which is considered a beneficial element to plants. Silicon may bring benefits to plants such as reduction of foliar diseases; improvement in pest control; increase in photosynthetic capacity due to the silicon benefit to the architectural activity of the plant, leaving the leaves more upright [32]; and improvement in the use of water by the plant [35]. Si may also influence the uptake and translocation of various macro- and micronutrients and increase plant tolerance to excess of Mn and Fe [37] and Zn, Al, and Cd [38].

Slag-based fertilizers applied in Si-deficient paddy soil improved rice growth, productivity, and brown spot resistance [39]. The application of steel slag in the soil increases the concentration of Si in the rice straw and promotes a higher yield of grains [40]. Also, studying the effect of slag in rice crop, [41] observed an increase of base saturation and availability of silicon and phosphorus in the soil, with a consequent increase in grain yield, Si content, and accumulation in rice straw, in 2 years of cultivation.

The supply of P and Si available in the soil for the potato crop, through the application of slag, increased the absorption of these nutrients by the plant, decreasing the lodging and increasing the height of plants and the production of tubers [17].

Nitrogen fertilization associated with steel slag also increases the dry mass production and the absorption of Si by the marandu grass [42].

Steel slag, used as soil acidity corrective material, increased the contents of Ca, Mg, P, and Si in the soybean leaves and Ca, Mg, and Si in maize and also provided increased shoot dry matter yield and grain yield of both cultures under no-tillage system; compared to limestone, steel slag was more effective in improving maize grain yield [27].

5. Conclusions

The steel industry generates large volumes of slag, which are considered as an environmental problem; it is necessary to increase the use of this waste in new processes to avoid disposal in landfills. Steel slags can be used in several activities, such as construction and paving, and also in the agricultural sector due to its ability to correct soil acidity, as it contains some nutrients for the plants and also as silicate fertilizer that is capable of providing silicon to the plants. Thus, steel slags can be considered as a sustainable alternative to agricultural practice.

Conflict of interest

There are no conflicts of interest to declare.

Author details

Angélica Cristina Fernandes Deus[1*], Rosemary Marques de Almeida Bertani[2], Guilherme Constantino Meirelles[1], Anelisa de Aquino Vidal Lacerda Soares[3], Lais Lorena Queiroz Moreira[4], Leonardo Theodoro Büll[1] and Dirceu Maximino Fernandes[1]

1 Department of Soil and Environmental Resources, College of Agronomic Sciences, São Paulo State University, Botucatu, São Paulo, Brazil

2 Paulista Agency Agribusiness Technology Regional Midwest Pole, Bauru, São Paulo, Brazil

3 Paulista Agency Agribusiness Technology, Regional Pole Midwest, UPD, Marilia, São Paulo, Brazil

4 Federal Institute of Techonology North of Minas Gerais, Januária, Minas Gerais, Brazil

*Address all correspondence to: angeldeys@hotmail.com

References

[1] Das B, Prakash S, Reddy PSR, Misra VN. An overview of utilization of slag and sludge from steel industries. Resources, Conservation and Recycling. 2007;**50**:40-57. DOI: 10.1016/j.resconrec.2006.05.008

[2] Alcarde JA, Rodella AA. Qualidade e legislação de fertilizantes e corretivos. In: Curi N, Marques JJ, Guilherme LRG, Lima JM, Lopes AS, Alvares VVH, editors. Tópicos em Ciência do Solo. Sociedade Brasileira de Ciência do Solo; 2003. pp. 291-334

[3] Babu T, Tubana B, Paye W, Kanke Y, Datnoff L. Establishing soil silicon test procedure and critical silicon level for rice in Louisiana soils. Communications in Soil Science and Plant Analysis. 2016: 1532-2416

[4] Yi H, Xu G, Cheng H, Wang J, Wan Y, Chen H. An overview of utilization of steel slag. Procedia Environmental Sciences. 2012;**16**:791-801. DOI: 10.1016/j.proenv.2012.10.108

[5] Shi C. Steel slag its production, processing, characteristics, and cementitious properties. Journal of Materials in Civil Engineering. 2004;**16**: 230-236. DOI: 10.1002/chin.200522249

[6] Torkashvand AM, Sedaghathoor S. Converter slag as a liming agent in the amelioration of acidic soils. International Journal of Agriculture and Biology. 2007;**9**:715-720

[7] Deus ACF. Aplicação de corretivos de acidez do solo na implantação do sistema plantio direto [thesis]. Botucatu: Faculdade de Ciências Agronômicas de Botucatu—Universidade Estadual Paulista Júlio de Mesquita Filho; 2014

[8] Alcarde JC, Rodella AA. Avaliação química de corretivos de acidez para fins agrícolas: uma nova proposição. Scientia Agricola. 1996;**53**:211-216. DOI: 10.1590/S0103-90161996000200003

[9] Brasil, Ministério da Agricultura. Secretaria Nacional de Defesa Agropecuária. Portaria No. 31, de 8 de junho de 1986. Determina as características físicas, PN e PRNT mínimas dos corretivos da acidez do solo: Classifica os calcários agrícolas em função do PRNT e determina como será calculado o PRNT. Diário Oficial, Brasília; 14 de junho de 1986. seção 1. p. 10.790

[10] Brasil, Instrução Normativa SDA no. 27, de 05 de junho de 2006 (Alterada pela IN SDA no. 7, de 12/04/2016, republicada em 02/05/2016). Os fertilizantes, corretivos, inoculantes e biofertilizantes, para serem produzidos, importados ou comercializados, deverão atender aos limites estabelecidos nesta Instrução Normativa no que se refere às concentrações máximas admitidas para agentes fitotóxicos, patogênicos ao homem, animais e plantas, metais pesados tóxicos, pragas e ervas daninhas. Brasília: Diário Oficial da República Federativa do Brasil; 2006

[11] Deus ACF, Bull LT, Correa JC, Villas Boas RL. Nutrient accumulation and biomass production of alfalfa after soil amendment with silicates. Revista Ceres. 2014;**61**:406-413. DOI: 10.1590/S0034-737X2014000300016

[12] Mantovani JR, Campos GM, Silva AB, Marques DJ, Putti FF, Langraf PRC, et al. Steel slag to correct soil acidity and as silicon source in coffee plants. African Journal of Agricultural Research. 2016;**11**:543-550. DOI: 10.5897/ajar2015.10535

[13] Souza RTX, Korndörfer GH. Slag efficacy as a lime and silicon source for rice crops trough the biological method. Journal of Plant Nutrition. 2010;**33**: 1103-1111

[14] Nolla A, Korndörfer GH, Silva CAT, Silva TRB, Zucarelli V, Silva MAG. Correcting soil acidity with the use of slags. African Journal of Agricultural Research. 2013;**8**:5174-5180. DOI: 10.5897/ajar2013.6940

[15] Alcarde JC. Corretivos da acidez do solo: Características e interpretações técnicas. São Paulo—SP: Associação nacional para difusão de adubos e corretivos (ANDA); 1992. 62 p. Boletim Técnico 6

[16] Prado RM, Fernandes FM, Natale W. Uso agrícola da escória de siderurgia no Brasil—Estudos na cultura da cana-de-açúcar. Jaboticabal, Funep. 2001. 68 p

[17] Pulz AL, Crusciol CAC, Lemos LB, Soratto RP. Influência de silicato e calcário na nutrição, produtividade e qualidade de batata sob deficiência hídrica. Revista Brasileira de Ciência do Solo. 2008;**32**:1651-1659

[18] Carvalho-Pupatto JG, Büll LT, Crusciol CAC. Atributos químicos do solo, crescimento radicular e produtividade do arroz de acordo com a aplicação de escórias. Pesquisa Agropecuária Brasileira. 2004;**39**: 1213-1218

[19] Corrêa JC, Büll LT, Crusciol CAC, Fernandes DM, Peres MGM. Aplicação superficial de diferentes fontes de corretivos no crescimento radicular e produtividade da aveia preta. Revista Brasileira de Ciência do Solo. 2008;**321**: 583-1590

[20] Corrêa JC, Büll LT, Crusciol CAC, Tecchio MC. Aplicação superficial de escória, lama cal, lodos de esgoto e calcário na cultura da soja. Pesquisa Agropecuária Brasileira. 2008;**43**: 1209-1219

[21] Deus ACF, Büll LT. Eficiência de escórias de siderurgia na cultura do feijoeiro em sistema de semeadura direta. Ciência Rural. 2013;**43**:1783-1789

[22] Correa JC, Freitag EE, Büll LT, Crusciol CAC, Fernandes DM, Marcelino R. Aplicação superficial de calcário e diferentes resíduos em soja cultivada no sistema plantio direto. Bragantia. 2009;**68**:1059-1068. DOI: 10.1590/S0006-87052009000400027

[23] Brassioli FB, Prado RM, Fernandes FM. Avaliação agronômica da escória de siderurgia na cana-de-açúcar durante cinco ciclos de produção. Bragantia. 2009;**68**:381-387

[24] Flower KC, Crabtree WL. Soil pH change after surface application of lime related to the levels of soil disturbance caused by no-tillage seeding machinery. Field Crops Research. 2011;**121**:75-87. DOI: 10.1016/j.fcr.2010.11.014

[25] Corrêa JC, Büll LT, Crusciol CAC, Marcelino R, Mauad M. Correção da acidez e mobilidade de íons em Latossolo com aplicação superficial de escória, lama cal, lodos de esgoto e calcário. Pesquisa Agropecuária Brasileira. 2007;**42**:1307-1317

[26] Castro GSA, Crusciol CAC. Effects of superficial liming and silicate application on soil fertility and crop yield under rotation. Geoderma. 2013; **195–196**:234-242

[27] Castro GSA, Crusciol CAC. Yield and mineral nutrition of soybean, maize, and Congo signal grass as affected by limestone and slag. Pesquisa Agropecuária Brasileira. 2013;**48**: 673-681. DOI: 10.1590/S0100-204X2013000600013

[28] Castro GSA, Crusciol CAC, Costa CHM, Ferrari Neto J, Mancuso MAC. Surface application of limestone and calcium-magnesium silicate in a tropical no-tillage system. Journal of Soil Science and Plant Nutrition. 2016;**16**:362-379.

DOI: 10.4067/S0718-951620160050 00034

[29] Pereira HS, Gama AJM, Camargo MS, Korndörfer GH. Reatividade de escórias silicatadas da indústria siderúrgica. Ciência e Agrotecnologia. 2010;**34**:382-390

[30] Prado RM, Fernandes FM. Efeito da escória de siderurgia e calcário na disponibilidade de fósforo de um Latossolo Vermelho-Amarelo cultivado com cana-deaçúcar. Pesquisa Agropecuária Brasileira. 2001;**36**: 1199-1204

[31] Amaral Sobrinho NMB, Velloso ACX, Oliveira C. Solubilidade de metais pesados em solo tratado com resíduo siderúrgico. Revista Brasileira de Ciência do Solo. 1997;**21**:9-16

[32] Corrêa JC, Büll LT, Paganini WS, Guerrini IA. Disponibilidade de metais pesados em Latossolo com aplicação superficial de escória, lama cal, lodos de esgoto e calcário. Pesquisa Agropecuária Brasileira. 2008;**43**:411-419

[33] Sposito G. The Chemistry of Soils. New York: Oxford University Press; 2008. p. 329

[34] Silveira MLA, Alleoni LRF, Guilherme LRG. Biosolids and heavy metals in soils. Scientia Agricola. 2003; **60**:793-806

[35] Deren CW, Datnoff LE, Snyder GH, Martin FG. Silicon concentration, disease response, and yield components of rice genotypes grown on flooded organic histosols. Crop Science. 1994;**34**: 733-737

[36] Korndörfer GH, Pereira HS, de Camargo MS. Silicatos de cálcio e magnésio na agricultura. Uberlândia: GPSi; 2003. p. 15. Boletim Técnico no 1

[37] Tavakkoli E, English P, Guppy CN. Interaction of silicon and phosphorus mitigate manganese toxicity in rice in a highly weathered soil. Communications in Soil Science and Plant Analysis. 2011; **42**:503-513

[38] Liang Y, Sun W, Zhu Y-G, Christie P. Mechanisms of silicon-mediated alleviation of abiotic stresses in higher plants: A review. Environmental Pollution, Barking. 2007;**147**:422-428

[39] Ning D, Song A, Fan F, Li Z, Liang Y. Effects of slag-based silicon fertilizer on rice growth and brown-spot resistance. PLoS One. 2014;**9**(7): e102681. DOI: 10.1371/journal. pone.0102681

[40] Ning D, Liang Y, Liu Z, Xiao J, Duan A. Impacts of steel-slag-based silicate fertilizer on soil acidity and silicon availability and metals-immobilization in a paddy soil. PLoS One. 2016;**11**(12):e0168163. DOI: 10.1371/journal.pone.0168163

[41] Barbosa Filho MP, Zimmermann FJP, Silva OF. Influence of calcium silicate slag on soil acidity and upland rice grain yield. Ciência agrotecnologia. 2004;**28**:323-331. DOI: 10.1590/ S1413-70542004000200011

[42] Fonseca IM, Prado RM, Vidal AA, Nogueira TAR. Efeito da escória, calcário e nitrogênio na absorção de silício e na produção de capim-marandu. Bragantia. 2009;**68**:221-232. DOI: 10.1590/S0006-87052009000100024

Chapter 5

Comprehensive Utilization of Iron-Bearing Converter Wastes

Hu Long, Dong Liu, Lie-Jun Li, Ming-Hua Bai, Yanzhong Jia and Wensheng Qiu

Abstract

Basic oxygen furnace (BOF) sludge is composed of not only valuable iron but also impurities like Zn, Pb, and some alkaline oxides. It is collected from wet cleaning system in steelmaking plants. How to deal with these double identity wastes? Will the traditional landfill treatments result in environmental pollution? What technologies have been developed recently, and is it actually useful? In this chapter, physical-chemical properties and mineralogical phases of converter sludge were characterized, and different recycling technologies were introduced. The proven metalized pellet-producing process would be highlighted that green pellets made from iron-bearing sludge are dried and preheated in a traveling grate firstly, and then reduced at high temperature in a rotary kiln or a rotary hearth furnace (RHF) to get direct reduced iron (DRI), served as a good iron source for blast furnace.

Keywords: BOF sludge, iron bearing, metalized pellet, direct reduced iron, environment friendly

1. Introduction

BOF sludge is collected from wet cleaning system in steelmaking plants. It is composed of not only valuable iron but also impurities like Zn, Pb, and some alkaline oxides [1–3]. Traditional landfill treatments inevitably result in environmental pollution because of the contained heavy metal and the high pH values of water-absorbed soil [4, 5]. Sintering is another recycling way to treat the sludge as raw material for sintering ore and fed to the blast furnace (BF). However, it leads to the circulation and accumulation of zinc in BF [6, 7]. To decrease the negative effects on BF production and surrounding environment, new recycling technologies have been developed, one proven of which is the metalized pellet-producing process, that green pellets made from iron-bearing sludge are dried and preheated in a traveling grate firstly, and then reduced at high temperature in a rotary kiln or a rotary hearth furnace (RHF) to get direct reduced iron (DRI), served as a good iron source for electric arc furnace (EAF) or BF [8–10]. It is an appropriate method to recycle the valuable iron and remove the harmful elements, simultaneously improve the burden structure of BF, which attracts more and more attention of metallurgical researchers.

Green pellets should be dried before charged into the consequent preheating and reduction process, or else they are easily to be cracked or pulverized owing to drastic volatilization of moisture at a high temperature, and results in the significant drop of the permeability, which will eventually lower the metallization rate of the products in the reduction furnace, or even make the production abnormal [11, 12]. Additionally, it has been reported that about 25% of energy for pellet induration is consumed for drying [12]. Thus, the improvement of the drying performance will be energy efficient, and also the drying mechanism of iron ore pellets made from BOF sludge is extremely important.

After dried and preheated, the mainstream approach dealing with pellets is baked first and then reduced by coal or reduction gas; another way is directly reduced so-called one-step method, one proven of which is the grate-rotary kiln process with high heat transfer efficiency, where the preheated iron-bearing material instead of fired oxide pellets directly reacts with the reductant of coal at high temperature to get direct reduced iron (DRI) [13, 14]. Compared with the traditional two-step rotary kiln, the new process is simplified and the risk of kiln accretion is reduced, because of the decreasing degradation without phase transferring from hematite to magnetite, which has been developed rapidly in China [15–17]. Metalized pellets could be used as burden for electric arc furnace (EAF) or BF, and the reduction degree, compressive strength and dezincification rate are considered as three important indicators [18–21]. They were energetically investigated by metallurgical researchers recently [9, 22, 23]. Many significant researches have been carried out to understand the reduction behavior of iron-bearing materials.

In this chapter, performances of pellets prepared from the BOF sludge are briefly presented at first and then studies on drying characteristics and reduction behavior under different conditions are introduced and compared to provide scientific guidance for recycling of secondary iron-bearing resource from BOF wastes [24–26].

2. Brief process of metallized pellet production

2.1 Two-step method

For traditional two-step method, oxidized pellets are prepared initially as raw materials for metallizing process, and then metallized pellets are produced through the direct reduction process, during which the temperature is below the melting point of the iron and the products referred to as direct reduced iron (DRI).

2.1.1 Preparation of oxidized pellets

The grate-rotary kiln producing line is considered and adopted as one of the most mature technologies, which has a strong adaptability to raw material and fuel, good quality, and low cost of production. The process consists of proportioning system, mixing, pelletizing system, green ball roller screen, and distribution, as well as grate-rotary kiln system for pellet indurations, finished product stock piling, and delivery system. The grate-kiln process flow is shown in **Figure 1**.

Concentrate fines are discharged to storage bins through the belt in the stockyard. Each concentrate store is equipped with disc feeder with frequency control and electric belt scale. Limestone and bentonite are delivered by tanker and then in a pneumatic transmission. Dust from multicyclone and electrostatic precipitators (ESP) for the main induced system is delivered by pneumatic transmission to dust bin. All the storage bins adopt weighing level gage to check material level in storage bin. The set value of charge ratio is automatically controlled and regulated by programmable logic controller (PLC) microprocessor. After all kinds of materials are compounded at a certain proportion, they are delivered directly to mixing room by the belt conveyor. Mixing room is usually equipped with a vertical mixer, which can mix materials both in micro- and macroway with high effectiveness, reliable operation, and simple maintenance. Materials mixed are

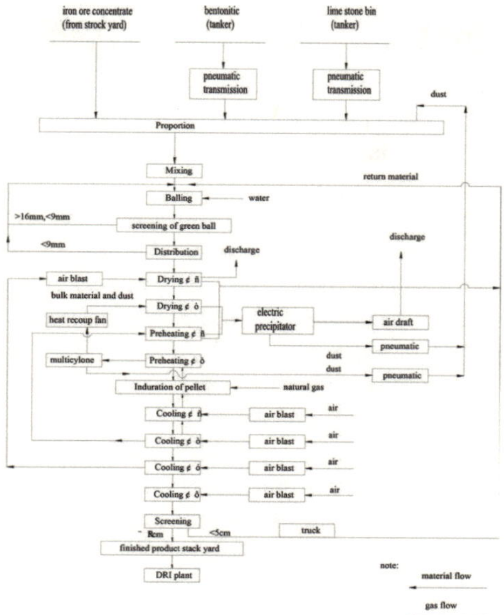

Figure 1.
Process flow chart of grate-kiln pelletizing plant.

delivered to pelletizing room by the belt conveyor. The belt conveyor is equipped with moisture detector for check and control of water addition for palletizing.

Disk pelletizers are often adopted for preparing green pellets. Revolving speed and inclination of disk are adjustable to guarantee the quality. Roller screener is equipped for screening, and pellets with suitable size are delivered to the distributer system in chain grate room, while others that unqualified are returned by the return conveyor system.

Chain grate area is mainly consisted of shuttle-type distributor, wide belt conveyor, roller distributer, chain grate, combustion burner fan for chain grate, electric double-dumping ash valve, bucket elevator and belt conveyor, etc. Shuttle distributor discharges green pellet to wide belt conveyor under motion back and forth, and the wide belt conveyor with speed in frequency control delivers green pellets to roller distributing device. Qualified green pellets are dried and preheated in the chain grate, and delivered to rotary kiln for roasting. Chain grate is divided into four zones, which are up-draft drying zone, down-draft drying zone, preheat I zone, and preheat II zone, respectively. The thickness of material bed in chain grate is about 180–200 mm, with the normal motion speed of 3.1 m/min. Pellets are first dried in the up-draft drying section by the recycle hot air in the temperature range of 160–250°C from the third cooling zone of annular cooler, which remove attached water in green pellets and to avoid pellets in the bottom of the grate bed from wetting. Temperature of hot returning gas from the annular cooler is in the range of 180–320°C, which can be mixed with cold air if necessary. In down-draft drying zone, 320–400°C recycle hot gas from the preheating II zone is pumped by hot-resistance draught fan across material layer from upper smoke shield, which makes green pellets dewater and dry, and can bear high-temperature stress and strain in 550–700°C in preheating I zone. The main induced draught fan is set to pump exhausted gas from air bellow, and eliminate it into atmosphere from electric precipitator. In preheating I zone, hot gas flow continues to dry green pellets through the material layer, and dried pellets begin to be oxidized, and the heat is derived from the hot exhausted gas of annular cooler II cooling zone. In preheating II zone, pellets are heated and oxidized further, and completely indurated, which makes pellets have a certain intensity, and can bear impact without breakage when falling to rotary kiln from chain grate in the motion of rotary kiln. Heat source comes from 900 to 1180°C hot air flow in rotary tail. A hole is left and sealed by fireclay brick in the partition wall between the first and second preheating zones. The hole can be opened in the case of thermal compensation. Hot exhausted gas in preheating I zone through collection header pipe of air bellow two sides and hot exhausted gas in down-draft drying zone are together discharged by electrostatic precipitator, main induced draught fan, and chimney. Chimney and valve are set at the top of the shield in preheating II zone, which is used for waste gas blowing off during furnace baking and emergency failure operation. Burners are installed in the cover of chain grate for thermal compensation. Preheated pellets get enough compressive, and then enter rotary kiln through kiln tail chute.

The rotary head is equipped with natural combustion apparatus with adjustable frame shape and length. Pellets are roasted while rotating in the kiln, so that the uniformity is assured. Baking temperature in rotary kiln is 1250–1320°C, revolving speed of rotary kiln is adjusted according to raw material differences to obtain enough retention time, and pellet quality. Liner made from precast brick and refractory castables makes kiln service with better thermal shock resistance and wear resistance and heat-shielding performance. An infrared simplified scanning temperature measurement system is adopted to detect and control the temperature inside. Roasted pellets are screened and discharged to receiving hopper of annular cooler.

Annular cooler consists of rotating part, air bellow, transmission device, rack, upper shield, and the system of variable-frequency adjustable-speed. It is mainly divided into four areas: I cooling zone, 900–1100°C of hot air is directly led to rotary kiln and used to raise temperature in kiln atmosphere; II cooling zone, 500–700°C of hot air returns to upper shield in preheating I zone of chain grate; III cooling zone, 180–320°C of hot air is led to up-draft drying section in grate; IV cooling section, 85–105°C of hot air is dedusted and exhausted through the chimney, with emission dust concentration below 50 mg/Nm^3. Fans are adopted to cool down pellets step by step and control the temperature of hot wind, and the majority of the heat generated during this cooling process is effectively reutilized. Pellets below 120°C are discharged through the hopper to belt conveyor, and then delivered out.

2.1.2 Preparation of metallized pellets

The attempts to develop large-scale direct processes have embraced practically every known type of apparatus suitable for the purpose including pot furnaces, reverberatory furnaces, shaft furnaces, rotary and stationary kilns, rotary hearth furnaces, electric furnaces, various combination furnaces, fluidized bed reactors, and plasma reactors. Many reducing agents including coal, coke, graphite, char, distillation residues, fuel oil, tar, producer gas, coal gas, water gas, and hydrogen have also been tried. For handling the BOF wastes, coal-based direct reduction technologies including the rotary kiln or the rotary hearth furnace (RHF) are often adopted in consideration of better material adaptability as well as for steady and reliable operation.

The representative two-step coal-based rotary kiln reduction process to produce sponge iron is given below including major technical parameters, types of equipment used, and the flow of materials through the plant. It is mainly composed of coal shed, iron raw material shed, coal screening building, proportioning building, rotary kiln-cooling building, etc. Schematic flowsheet of this rotary-kiln sponge iron plant is shown in **Figure 2** [27].

The lump ore or pellet is transported into DRI plant by dump trucks. The iron raw material feeds the belt conveyor by wheel loader then delivered to the proportioning building. The reductant (coal) is transported into DRI plant by dump trucks. The

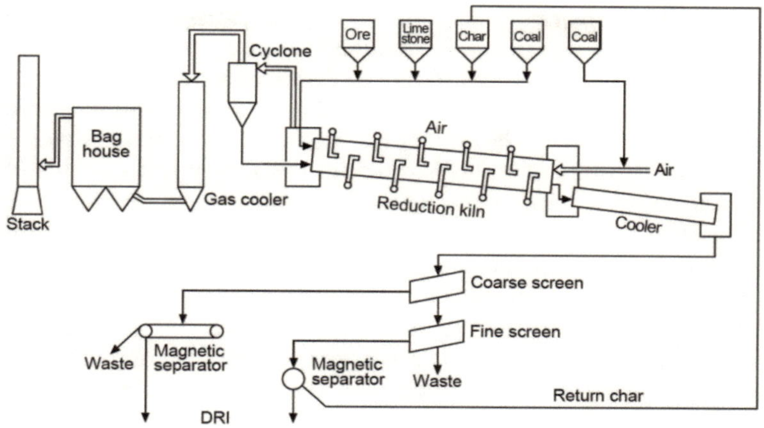

Figure 2.
Schematic flowsheet of rotary-kiln sponge iron process.

capacity of the reductant (coal) storing yard is for about 16 days. Coal is fed into the belt conveyor by wheel loader and then delivered to the coal screening building for classification. The grain size of reductant is 0–50 mm. According to the process requirement, the grain size of the coal fed at the head of rotary kiln is 3–25 mm and at the tail is 25–35 mm. The coal is separated into four granularities: 0–3, 3–25, 25–35, and 35–50 mm by linear screen. The coal with grain size 3–25 mm will be delivered to the coal hopper at the head of the rotary kiln, coal with grain size 25–35 mm is delivered to the coal bin in the proportioning building by belt conveyor, and coals with grain sizes 0–3 and 35–50 mm are delivered to power plant by truck. The desulfurizer (limestone) is transported into DRI plant by dump trucks and discharged into the receiving hopper in the proportioning building, and then delivered to the limestone bins in the proportioning building by steeply belt conveyor. The proportioning of iron raw material, limestone, and coal is completed in the proportioning building. The belt scale is adopted to make proportioning accurate.

The rotary kiln is one of the main equipments used to produce directly reduced iron products. The mixed material fed into the rotary kiln is heated up to a certain temperature in the kiln, and then the reduction reaction will take place. Detailed process is as follows. The iron raw materials, coal, and limestone used for reduction are fed at the head and tail of the rotary kiln separately. The granular coal injection gunner and the ignitor based on diesel oil will be settled in the head of rotary kiln. The diesel oil ignitor is set to increase temperature up to the range of 600–800°C. The grain size of coal fed at the head is 3–25 mm, which is injected to the rotary kiln by granular coal injection gunner. The grain size of coal fed into the rotary kiln at the tail is 25–35 mm, together with limestone and iron raw material by belt scale. According to the chemical equation $C + CO_2 = 2CO$, the CO will react with the oxygen contained in the pellet. The rest of the CO is burnt with the secondary injecting hot air to heat up the rotary kiln and the pellets. A series of chemical reactions take place with the iron raw materials, reductant, and desulfurizer, and then the oxidized pellets are reduced to metallized pellets, so-called sponge iron. The sponge iron is discharged to the cooler drum from the end of the rotary kiln. The temperature inside the rotary kiln is in the range of 1000–1100°C, and the retention time inside is about 5–9 h. After that, the sponge iron at high temperature is discharged to the fixed screen and sent to the cooler drum to be cooled.

The cooler drum is self-sealed with credible sealing device at the feed end and discharge end, and operated with micropositive pressure with the inner part isolated from the outside environment (air). To avoid oxidation, the product is cooled down below 120°C indirectly by spraying water to the outside surface of the cooler drum. The material will stay inside for about 25–40 min.

The cooled product is delivered to the product sorting room to screening and classification in order to separate the DRI, magnetic powder, nonmagnetic powder, and return coal fines. The product discharged from the cooler drum is delivered to the screen equipment and separated into two granularities: >4 and <4 mm. To minimize the iron content in nonmagnetic powder, the product of >4 mm will be proceed in three steps in magnetic separator. The magnetic material discharged from the magnetic separator is fed to the steel melt shop. The nonmagnetic material is discharged from the bottom of magnetic separator, regarded as return coal, and delivered to proportioning building. The product of <4 mm is separated in two serial elutriator for pneumatic classification: the light one such as ash is dedusted, and the heavier one is pneumatic delivered to the single magnetic separator. After magnetic classification, the nonmagnetic powder is delivered to nonmagnetic bin by belt conveyor and bucket elevator and then transferred by trucks; the magnetic material will be separated in elutriator, delivered to magnetic bin by belt conveyor and then pressed into block, and finally delivered to the steel melt shop too.

2.2 One-step method

The “one-step” method is mainly composed of concentrate pelletizing—preheating—direct reduction (coal-based kiln reduction/rotary hearth furnace)—cooling. Compared with the traditional two-step method, the preheated pellets with certain intensity are directly delivered to the direct reduction furnace, which not only shortens the process line with satisfactory product quality, but also saves the energy.

The key equipment reform and technologies of one-step kiln process are primarily researched and applied into operation in China. Reduction behavior of two kinds of pellets using noncoking coal as reductant is studied and compared. The one is preheated pellet made of magnetite concentrate with composite binder, and the other is fired oxide pellets containing bentonite as binder [15]. Reducibility, compress strength, porosity, and microphases evolution are measured. Results show that preheated pellets possess much better reducibility than fired oxide pellets, which is ascribed to their higher effective diffusivity due to higher porosity. The compressive strength increases obviously after 30 min reduction and achieves a high value at the end of reduction, while the value of metalized pellets from reduction of oxidized pellets is much lower, because of more cracks and fractures formed. Xinjiang magnetite concentrate in China was studied by one-step process for direct reduction. The results show that for Xinjiang magnetite assaying with 69.21% Fe, damp milling and adding agent can improve the quality of green balls obviously. After drying, green balls are preheated at 800°C for 10 min, pellets with compressive strength of 581 N/pellet are achieved, preheated pellets were directly reduced at 1050°C for 80 min, and directly reduced iron assaying 90.33% of total iron and 85.05% of metallic iron and 94.15% of metallization degree are achieved. Compared with the traditional direct reduction of fired oxide pellets in coal-based rotary kilns, one-step process for direct reduction can avoid high-temperature oxidation of pellets at 1150–1300°C, and possess some advantages such as greater economic profit and good quality of direct reduced iron [28]. Weike one-step DRI plant with the annual output of 62,000 ton designed by Changsha Metallurgical Design and Research Institute in China started production in 2002, after more than 2 years trial test and technology renovation [29].

The FASTMET process, developed during the early 1990s by Midrex Direct Reduction Corporation to provide a coal-based process for North American locations, is very similar to the rotary hearth pioneered by Inmetco in Ellwood City,

Pennsylvania for treating waste dusts from steel plants and also a proposed process studied by Salem Furnace Company in Carnegie, Pennsylvania. The nucleus of the FASTMET process originated with the development of the heat fast process in the mid-1960s. The heat fast process consisted of the following steps: (1) mixing and pelletization of iron ore fines and pulverized coal, (2) drying the green pellets on a grate, (3) prereduction on an RHF, and (4) cooling in a shaft furnace. The next step in the evolution of the FASTMET process was the development of the Inmetco process for reduction of stainless steel mill wastes in 1974. The process consisted of the following steps: (1) mixing and pelletization of mill wastes and pulverized coal, (2) prereduction of the green pellets on an RHF, (3) discharge of hot pellets into transfer bins, and (4) charging of hot pellets into a submerged arc furnace. The process was tested in a pilot plant located at Port Colborne, Ontario, and a commercial unit of 60,000 tons per year capacity was set up in 1978 at Ellwood City, Pennsylvania.

Midrex revived investigations in 1989 and tested a wide range of raw materials in its laboratory in Charlotte, North Carolina. A 2.5 m (8.2 ft) diameter pilot RHF with a capacity of 150 kg/h of reduced iron was installed in 1991 to simulate the reduction portion of the process. This process simulator was utilized to develop data for the design of an industrial unit. Based on the successful laboratory and pilot plant tests, Midrex and its parent company Kobe Steel started construction of an FASTMET demonstration plant in April 1994 at the Kakogawa works of Kobe Steel in Japan. The plant has a production capacity of 2.6 tones/h and was commissioned in September 1995. The demonstration plant is reportedly operating under stable conditions, and several tests have been conducted to develop process parameters for scale-up to a 60 tone/h industrial scale unit. Tests have also been made for producing hot briquetted iron (HBI) and for integrating the FASTMET process with DRI melting in an electric arc furnace.

In the FASTMET process, shown schematically in **Figure 3**, iron ore concentrate, reductant, and binder are mixed and formed into green pellets that are dried at 120°C and fed to the rotary hearth furnace. The pellets are placed on a solid hearth one to two layers deep as shown in **Figure 4** [30–32].

As the hearth rotates, the pellets are heated to 1250–1350°C by means of fuel burners firing into the freeboard above the hearth. Reduction to metallic iron is completed in 6–12 min, depending on the materials, temperature, and other

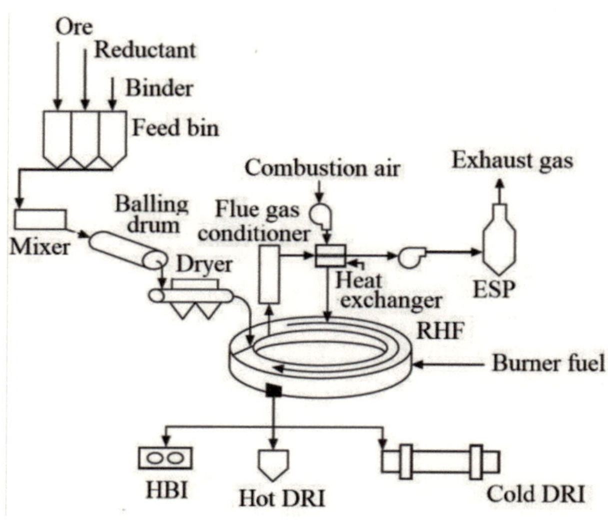

Figure 3.
Schematic flowsheet of FASTMET process [30].

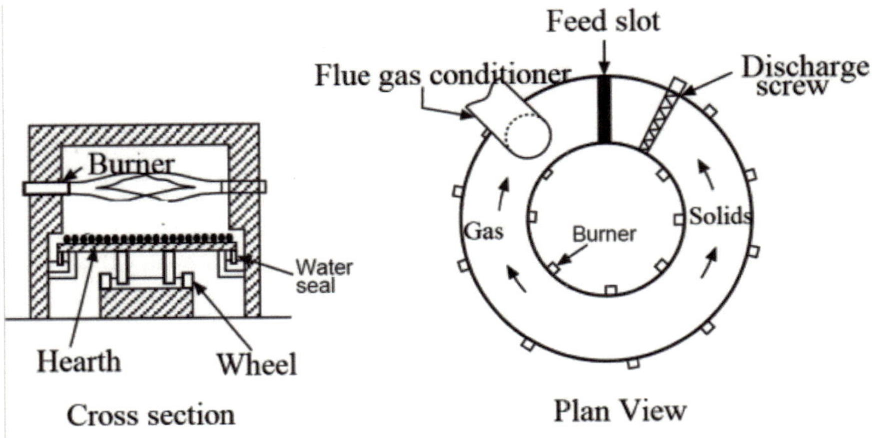

Figure 4.
Cross section and plan view of the rotary hearth furnace [30].

factors. The DRI is discharged at approximately 1000°C and can be hot charged to an adjacent melter, hot briquetted, or cooled indirectly before storage and/or shipment. The residence time of the pellets in the ?RHF is 12 min. The hot, reduced iron from the RHF is partially cooled to about 1000°C by a water-cooled discharge screw. The product can be obtained either in the form of cold DRI, hot DRI, or HBI. The product can be discharged from the RHF either to a transfer device for conveying the hot DRI directly to an adjacent steelmaking facility, or by gravity to a briquetting system for producing HBI. The off-gas from the RHF flows to a gas handling system, where the SO_2, NO_x, and particulates can be reduced to the desired limits. The hot off-gas is used for preheating the combustion air for the RHF burners and to supply the heat for drying. The process is designed to recycle all process water. All fines generated in the process are recycled though the feed.

3. Properties of pellets made from BOF wastes

General metallized pellet production processes are introduced above. In this section, the detailed properties of pellets made from BOF wastes, drying, and reduction behavior will be set forth further.

3.1 Generation and properties of BOF wastes

Converter sludge is a kind of black slurry with a high water content. It becomes a dense lump after dehydration and then dispersed into fine particles with large specific surface area after dispersion, among which the content of particle with a size below 0.075 mm is larger than 70%, while the percentage below 0.048 mm is larger than 50%. As the dust and mud are so fine that the surface activity is relatively large, it is easy to be absorbed and blown up into the air after drying, which seriously pollutes the surrounding environment. In terms of chemical composition, the total iron content is high and the impurity content is low. Most of them have simple composition, high content of iron ore, and relatively few impurities, which is conducive to comprehensive recovery and utilization. However, a strong alkaline hydroxide would like to be formed after absorbing water because of the contained CaO, MgO, K_2O, and Na_2O, which may lead to the increase of pH of the water and soil around and is bad for the growth of crops. Hence, reasonable recycling of BOF wastes is very important.

Item	TFe	FeO	CaO	MgO	SiO_2	MnO	ZnO	Al_2O_3	PbO	K_2O	Na_2O
wt, %	54.53	58.54	16.38	5.75	2.31	0.9	0.77	0.74	0.36	0.18	0.1

Table 1.
Major chemical composition of converter sludge.

Chemical composition of the converter sludge obtained from a steelmaking company is shown in **Table 1**. Then, mineralogical phase was analyzed through the Laitz DMRX polarization microscope and the mineralogical content was characterized through X-ray diffraction (XRD) analysis.

3.2 Properties of pellets prepared from BOF sludge

To investigate the properties of pellets made from BOF sludge, the raw material was dried (at 105°C for 24 h in a drying oven) and finely ground with the size distribution listed in **Table 2**. Then, they were continuously charged into a pelletizing disk to prepare the green pellets. After that, the initial moisture content was obtained through drying method. In total, 100 g of green pellets were weighed and dried in the oven at 200°C for 24 h, and the difference of weights before and after drying was calculated as the moisture content. The compressive strength was measured according to the standard of ISO470016, and the falling strength (drop no.) was counted through repeated pellet falling from a height of 0.5 m to the steel plate with the thickness of 0.01 m until it cracked. In total, 10 balls were measured and the average was taken.

The bursting temperature was defined as the maximum temperature at which bursting rate was less than 4%. In total, 50 pellets were placed in a baking cup and heated for 3 min at the predetermined temperature with the flow rate of 1.5 m/s, then moved out, and observed. If two of them were cracked, the relative temperature was recorded as a bursting temperature.

Properties of green pellets were listed in **Table 3**. Moisture content was in the range of 15.29–16.78%, which was much higher compared to ordinary iron ore pellets (about 7–9%) [17, 18], and would make the drying more difficult. Pellets were strong enough with both high compressive strength and falling strength. However, the bursting temperature was low, so the temperature in the primary drying should be set as 150°C to avoid great bursting.

Figure 5 shows optical microstructures of the converter sludge. The major iron-bearing phases are magnetite (Fe_3O_4), iron, and wustite ($Fe_{1-x}O$) (**Figure 5a** and **b**). Some of hematite (Fe_2O_3) (**Figure 5a**) is also observed. Most of the iron-bearing phases are spherical, with the grain size smaller than 50 μm. Gangues observed are mainly composed of silicate crystal phase and glass phase, as shown in **Figure 5b** and **c**, and also coke particles are found in **Figure 5c** and **d**.

Size, mm	>0.5	0.5–0.3	0.3–0.15	0.15–0.105	0.105–0.076	<0.076
wt, %	36.74	8.46	11.24	1.33	14.33	27.90

Table 2.
Size distribution of converter sludge after ground.

Item	Moisture (wt, %)	Compressive strength (N/p)	Falling strength (no. of drops/p)	Bursting temperature (°C)
Value	15.29–16.78	11.27–16.27	>20	150

Table 3.
Properties of green pellets made from converter sludge.

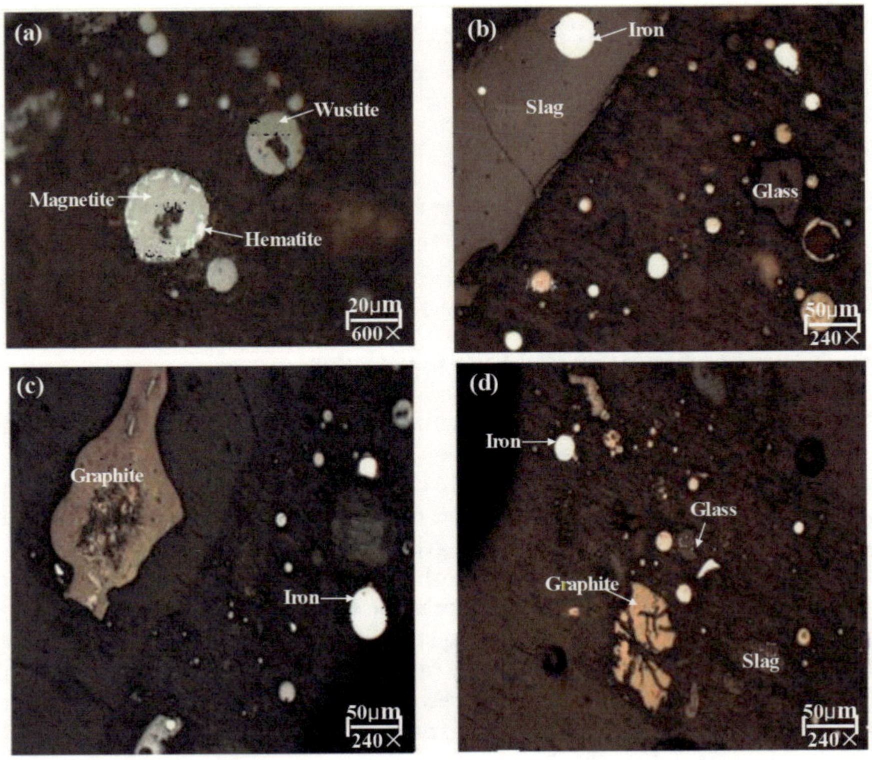

Figure 5.
Optical microstructures of the converter sludge.

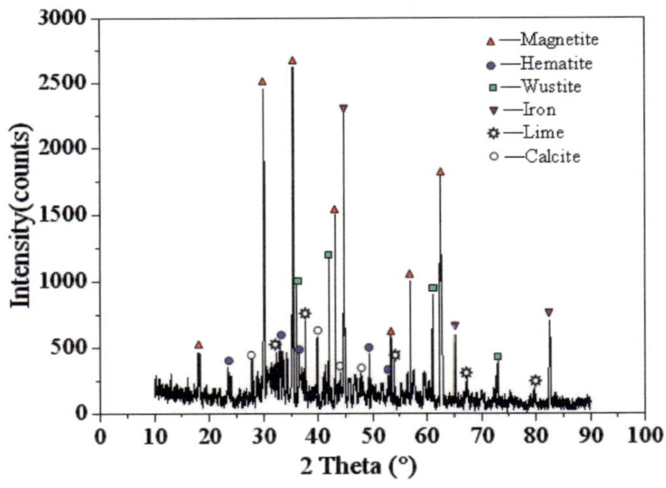

Figure 6.
XRD analysis of the converter sludge.

This mineralogical composition of converter sludge is verified through XRD analysis shown in **Figure 6**. The particularly intense peaks of magnetite, iron, and wustite prove that they mainly comprise iron contained phase. The content of hematite is relatively lower. For the slag phases observed above, they are predominantly identified as CaO and $CaCO_3$.

The $CaCO_3$ is formed during the pelletizing with the reaction of formed slaking $Ca(OH)_2$ and CO_2 in the air, and two positive charges of Ca^{2+} will produce strong electrostatic adsorption, which makes the green pellets strong enough [33]. However, so much lime in the sludge also brings the issue of much moisture content in the pellet (~16.52% in average) because of its high hydrophilic ability, which directly increases the risk of bursting during the drying.

4. Drying mechanism of BOF sludge pellets

Green pellets should be dried before being charged into the consequent preheating and reduction process or else they are easily to be cracked or pulverized owing to drastic volatilization of moisture at a high temperature, and this results in the significant drop of the permeability, which will eventually lower the metallization rate of the products in the reduction furnace, or even make the production abnormal. Additionally, it has been reported that about 25% of energy for pellet induration is consumed for drying [12]. Thus, the improvement of the drying performance will be energy efficient. The drying mechanism of iron ore pellets has been studied by several researchers [12, 34–36]. Two-stage drying is assumed for most drying models at first, involving surface evaporation and internal drying after a critical moisture content is reached [12]. Then, four-step drying kinetics of individual pellet is proposed and verified through experimental and numerical research by Tsukerman et al. [34]. Feng et al. investigate effects of some parameters on drying properties [35, 36].

Further research about drying characteristics of converter sludge pellets was studied to promote its recycle and reutilization [36]. Influence of factors including temperature, time, and flow on their drying properties were studied and optimized through experiments in order to get better reuse of secondary resources through the grate-kiln (or RHF) metalized pellet-producing process. Following is a brief introduction about the experiments.

In an actual production, drying in the traveling grate was divided into two stages [37]. This drying test was performed in the baking cup (with diameter of 80 mm and length of 300 mm) equipped with a crossflow adjusting system to simulate the two-stage thermal condition. The schematic was shown in **Figure 7**. Pellets are divided into three layers and placed in the baking cup (each layer with the height of 60–70 mm, and separated by a meshed stainless steel plate). Afterward, they were quickly covered by the burner and dried according to the scheduled time, flow, and temperature. Finally, they were moved out and cooled naturally. The bursting pellets for each layer were collected and weighed, respectively, and the remaining moisture was measured using the same method as the green pellets.

An orthogonal array was applied in this drying test. Four technical parameters were selected based on operation experience in pelletizing plant: first drying stage time (t_1: 5, 7, 9 min), the second drying temperature (T_2: 200, 300, 400°C), second drying time (t_2: 2, 4, 6 min), and second drying flow rate (v_2: 1.0, 1.4, 1.8 m/s, standard state), respectively. The temperature and flow rate in the first drying stage were set as fixed factors (T_1 = 150°C; v_1 = 1.5 m/s) according to the previous bursting temperature measurement. Bursting rate of pellets and the dehydration rate were considered as two major indicators of drying characteristics and calculated according to the following equations.

$$\text{Busting rate (\%)} = m_1/m_0 \times 100 \tag{1}$$

$$\text{Dehydration rate (\%)} = (m_2 - m_3)/m_2 \times 100 \quad (2)$$

where, m_1 is the weight of bursting pellets, m_0 is the weight of total pellets, m_2 is the weight of water in green pellets, and m_3 is the weight of water in dried pellets.

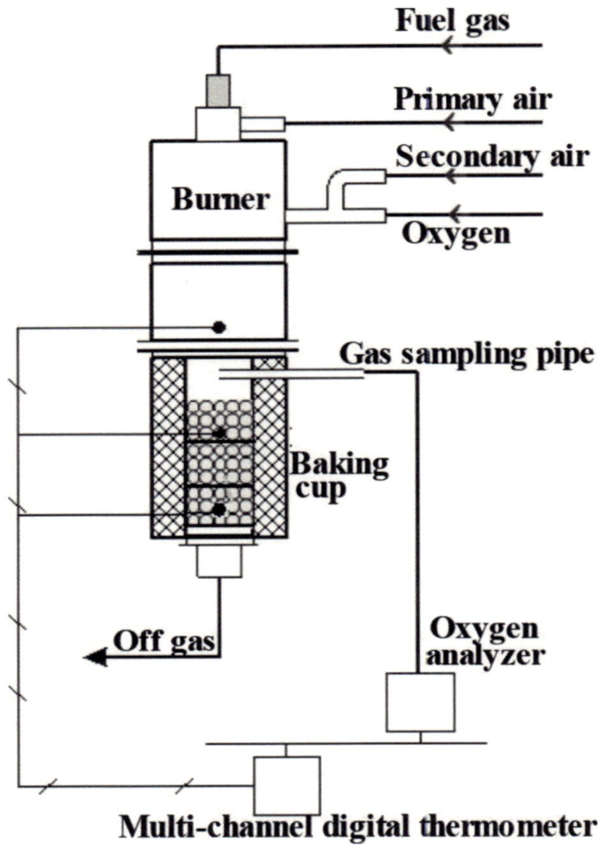

Figure 7.
Schematic of the drying simulator.

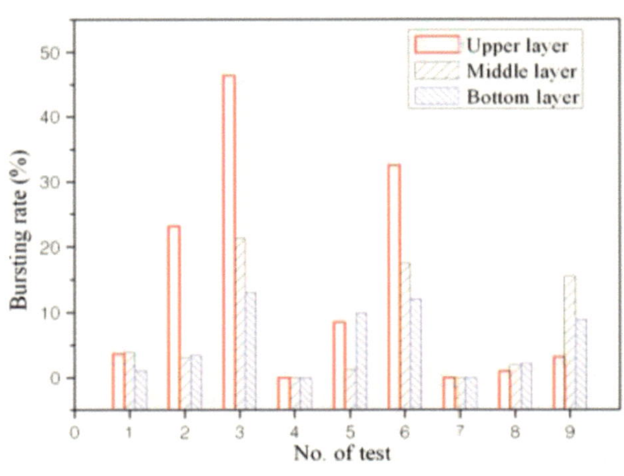

Figure 8.
Bursting rate of dried pellets.

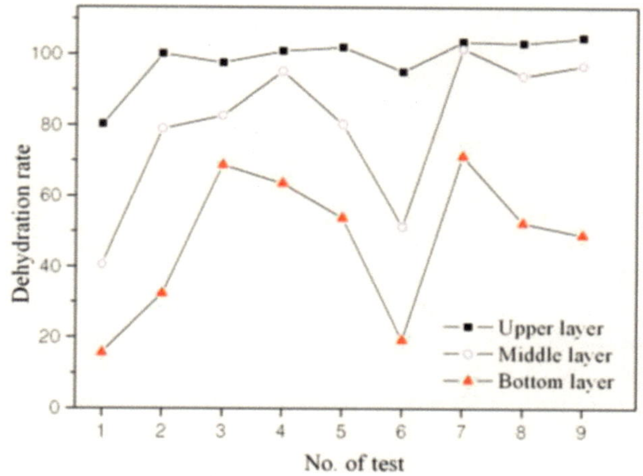

Figure 9.
Dehydration rate of dried pellets.

Figure 8 shows the bursting rate of pellets in each layer for the nine tests. High bursting rate means poor drying performance. It is found that bursting rate of the upper pellets in nos. 3, 6, and 2 is the worst, with the values of 46.41, 32.47, and 23.12%, respectively. Common conditions they share are short drying time (5–7 min) in first drying section, high level of temperature (300–400°C), and fast flow rate (1.4–1.8 m/s) in second drying section. It can be speculated that if the first drying time is not long enough, cracking rate of pellets will go up during the second drying section caused by the excessive expansion force generated through rapid and intense vaporization of residual moisture in the interior of pellet at high temperature and flow rate.

Figure 9 shows the dehydration rate of pellets in each layer. Obviously, the bottom pellets have the lower dehydration percentage compared with the upper and middle one, the lowest value of which are nos.1, 2, and 6, with 15.58, 32.31, and 19.16%, respectively.

Actually, moisture evaporation is a dynamic balance of liquid water gasification and vapor condensation, and evaporation rate is the difference of these two reactions [25]. It follows that the moisture evaporation velocity depends on the temperature and the pressure difference between saturated and ambient vapor. The higher the temperature and pressure difference, the faster is the evaporation rate. So, this is the reason why the dehydration rate in the bottom level is the lowest. For one thing, the temperature at the bottom is lower than the upper one as the endothermic evaporation reaction takes place when gas flows from up to down. For another, the vapor generated in the upper gradually enters into the main flow, which will lower the differential pressure of vapor.

5. Reduction mechanism of BOF sludge pellets

Most DRI production is melted in electric arc furnaces for steelmaking. Minor amounts may be charged into the ironmaking blast furnace. So the reduction degree, compressive strength, and dezincification rate are considered as three important indicators. They were energetically investigated by metallurgical researchers recently. The study of reduction behavior of the composite briquettes shows that both the metallization and dezincification ratios increased with the increasing temperature and the time, but first increased and then decreased with the increasing C/O molar ratio [9]. The strength of the metallized pellet could be controlled by reduction temperature, sintering time, additive quality/quantity, and manner of reduction [22]. DRI strength values are found to decrease from 200 to 30 kg during reduction and then strengthened up to 50 kg as a result of sintering and fusion [23].

In order to clarify reduction mechanism of BOF sludge pellets, experiments to simulate the grate-kiln metalizing process is conducted. Flow diagram is shown in **Figure 10**, including converter sludge pellets preparing, drying, preheating in the baking cup, and subsequent direct reduction in the preheated pellets with reductant of coal in the simulated rotary kiln.

Experimental materials include the iron-bearing converter sludge and coal. Converter sludge's chemical composition and mineralogical phase are shown above. It mainly contains 54.53% TFe, 16.38% CaO, and 0.77% ZnO (weight percentage). The major iron-bearing phases are magnetite (Fe_3O_4), iron, and wustite ($Fe_{1-x}O$). Chemical analysis of coal as the reductant and its softening and melting

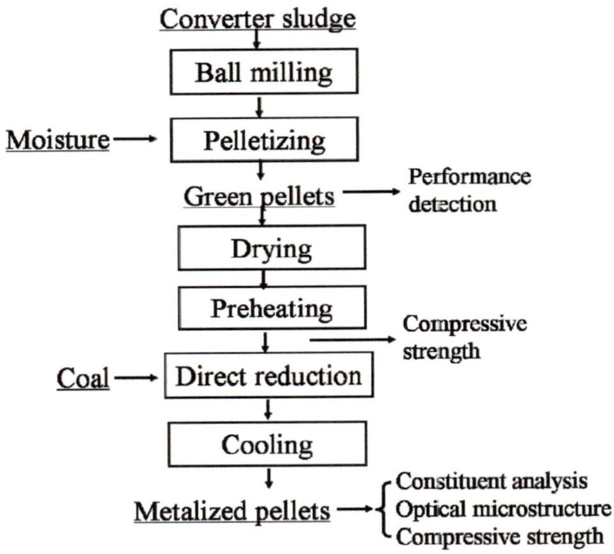

Figure 10.
Flow diagram of direct reduction experiment.

Composition	FC, ad	M, ad	A, ad	V, ad	V, daf
wt, %	76.18	0.72	9.68	13.42	14.98

ad = on air dry basis, daf = on dry ash free, FC = fixed carbon, M = moisture, A = ash, and V = volatile.

Table 4.
Chemical analysis of coal.

Item	Distortional temperature	Softening temperature	Half global temperature	Flowing temperature
Value, °C	1125	1180	1210	1280

Table 5.
Softening and melting properties of coal.

properties are shown in **Tables 4** and **5**, indicating it is suitable for direct reduction with high ash melting point.

The simulated rotary kiln is shown in **Figure 11**. It mainly consisted of a rotary drum (with a diameter of 130 mm and a length of 200 mm, made of heat-resistant steel) and an electrically heated tube furnace.

About 500 g of preheated pellets and coal at different ratio (C/O molar ratio) were put into the steel drum, which had been heated up to the target temperature in the furnace. Afterward, the direct reduction is proceeded for the predetermined residence time in the drum with a rotation speed of 30 rpm. Finally, they were moved out and cooled down to the ambient temperature. Optical microstructures of metallized pellets were analyzed through microscope. The compressive strengths before and after reduction were measured. Metallization rate was calculated on basis of chemical analysis according to the following equations.

$$\text{Metallization rate (\%)} = \text{MFe(\%)/TFe(\%)} \times 100 \tag{3}$$

where, MFe means metallic iron, and TFe means total iron.

Nine tests were conducted, and three technical parameters including temperature (1000, 1050, and 1100°C), time (1.5, 2.0, 2.5 h), and coal ratio (C/O molar ratio, 1.1, 1.3, and 1.5) were selected. Results show that the nos. 3 (T = 1050°C; t = 2.5 h; C/O = 1.5), 2 (T = 1050°C; t = 2 h; C/O = 1.3), and 6 tests (T = 1100°C; t = 2.5 h; C/O = 1.1) have the highest metallization rate of 74.7, 47.7, 45.7%, respectively. Common conditions they share are relatively long reaction time and high reduction temperature. Their corresponding compressive strengths are 1014, 897.4 and 1506.7 N/p, indicating pellets produced from certain tests meet the strength requirement of material served for BF. In addition, the residual zinc contents of these three reduced pellets are 0.44, 0.54, and 0.38%, and the average dezincification rate is calculated as 41.6%. This index can be further improved in an actual rotary kiln because the vapor of zinc produced during reduction could be brought out with the flow.

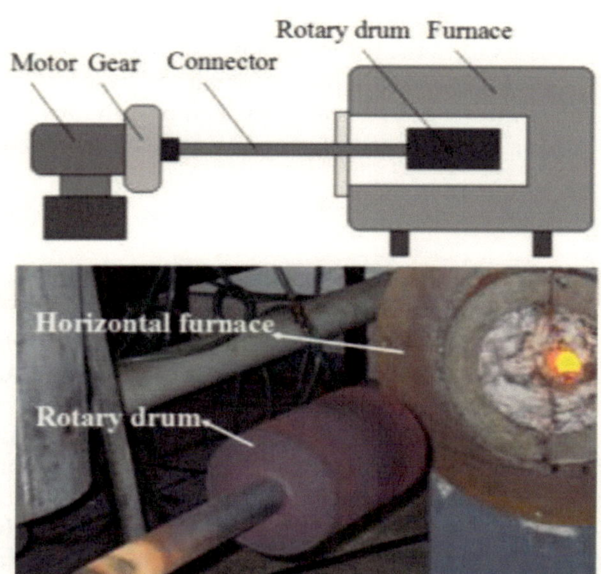

Figure 11.
Schematic of the metalizing simulator.

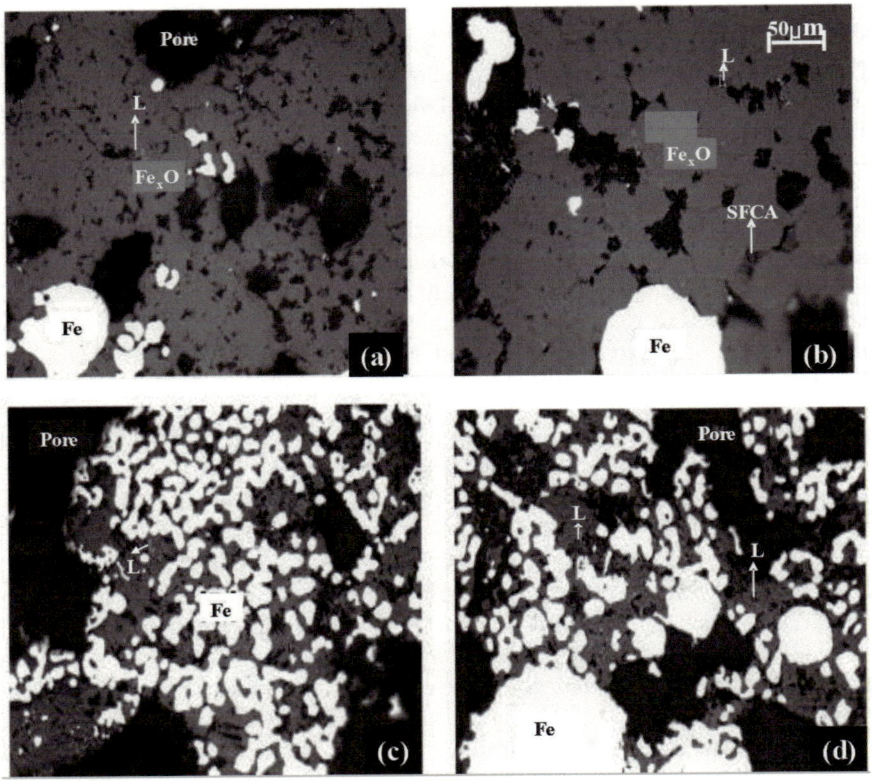

Figure 12.
Optical reflecting microstructures of the reduced pellets, 200×: (a) and (b) interior and (c) and (d) exterior. L: silicate, SFCA: composite calcium ferrite, Fe: metallic iron, and Fe_xO: wustite.

Figure 12 shows optical microstructures of the reduced pellets. A lot of wustite exists in the interior of metalized pellets as shown in **Figure 12a** and **b**, while a large amount of metallic iron, which looks much whiter and brighter than wustite, is mainly observed in the exterior shown in **Figure 12c** and **d**. It can be deduced that the reduction condition in the out part of pellets is much better than the interior at the beginning, and the compact shell of iron rapidly formed in the initial makes the diffusion of reduction gas more difficult from outside to the inner.

6. Conclusions

Converter sludge is a kind of useful secondary resource rich in valuable iron and calcium oxide. High initial moisture in converter sludge enhanced the bursting risk during drying and made this process difficult. The upper layer bursting rate and the bottom layer dehydration rate are considered as main indicators of drying performance. Technical factors of temperature and retention time in the drying section have remarkable influence on the drying performance, which should be paid close attention in future.

Two-step and one-step coal-based methods have been extensively researched recently and adopted into operation of the BOF wastes recycling. Effects of direct reduction parameters including temperature, time, and coal ratio on the metallization rate and compressive strength are also studied. Results show that reduction time and temperature have remarkable influence on the metallization rate. The indexes of metallization rate and compressive strength of metallized pellets reduced from the BOF sludge can satisfy the requirement of iron burden for BF. To improve the efficiency, the gas-based direct reduction especially hydrogen as the reductants may be considered to develop breakthrough technologies for emission reduction.

Acknowledgements

The financial support from Natural Science Foundation of Hebei Province in China (No. E2017203157) and China Postdoctoral Science Foundation is gratefully acknowledged. The authors also thank Professor Yanzhong Jia and Delan Liang (University of Science and Technology Beijing) for their supports.

Conflict of interest

There is no conflict of interest.

Other declarations

Mutual development of metallurgical technology and ecological environment is our persistent pursuit.

Author details

Hu Long[1*], Dong Liu[2], Lie-Jun Li[1], Ming-Hua Bai[3], Yanzhong Jia[4] and Wensheng Qiu[2]

1 School of Mechanical and Automotive Engineering, South China University of Technology, Guangzhou, Guangdong, P. R. China

2 Baowu Group Guangdong Shaoguan Iron and Steel Co., Ltd., Shaoguan, Guangdong, P. R. China

3 National Engineering Research Center for Equipment and Technology of Cold Strip Rolling, Yanshan University, Qinhuangdao, Hebei, P. R. China

4 School of Metallurgical and Ecological Engineering, University of Science and Technology Beijing, Beijing, P. R. China

*Address all correspondence to: longhu042@126.com

References

[1] Cristina N, Lobato C, Villegas EA, Mansur MB. Management of solid wastes from steelmaking and galvanizing processes: A brief review. Resources, Conservation and Recycling. 2015;**102**:49-57

[2] She XF, Xue QG, Dong JJ, Wang JS, Zeng H, Li HF, et al. Study on basic properties of typical industrial dust from iron and steel plant and analysis of its utilization. The Chinese Journal of Process Engineering. 2009;**9**:7-12 (in Chinese)

[3] Zhang YB, Liu BB, Xiong L, Li GH, Jiang T. Recycling of carbonaceous iron-bearing dusts from iron & steel plants by composite agglomeration process (CAP). Ironmaking and Steelmaking. 2017;**44**:532-543

[4] Das B, Prakash S, Reddy PSR, Misra VN. An overview of utilization of slag and sludge from steel industries. Resources, Conservation and Recycling. 2007;**50**(1):40-57

[5] López FA, Balcázar N, Formoso A, Pinto M, Rodriguez M. The recycling of Linz-Donawitz (LD) converter slag by use as a liming agent on pasture land. Waste Management and Research. 1995;**13**(6):555-568

[6] Makkonen HT, Heino J, Laitila L, et al. Optimisation of steel plant recycling in Finland: Dusts, scales and sludge. Resources, Conservation and Recycling. 2002;**35**(1-2):77-84

[7] Besta P, Samolejová A, Janovská K, Lenort R, Haverland J. Utjecaj štetnih elemenata pri proizvodnji sirovog željeza u odnosu na ulaznu i izlaznu materijalnu bilancu. Metalurgija. 2012;**51**(3):325-328

[8] Mombellia D, Di Cecca C, Mapelli C, Barella S, Bondi E. Experimental analysis on the use of BF-sludge for the reduction of BOF-powders to direct reduced iron (DRI) production. Process Safety and Environmental Protection. 2016;**102**:410-420

[9] Xia LG, Mao R, Zhang JL, Xu XN, Wei MF, Yang FH. Reduction process and zinc removal from composite briquettes composed of dust and sludge from a steel enterprise. International Journal of Minerals, Metallurgy, and Materials. 2015;**22**(2):122-131

[10] Su FW, Lampinen HO, Robinson R. Recycling of sludge and dust to the BOF converter by cold bonded pelletizing. ISIJ International. 2004;**44**(4):770-776

[11] Prieto Martinez N, Herrera Trejo M, Morales Estrella R, et al. Induration process of pellets prepared from mixed magnetite–35% hematite concentrates. ISIJ International. 2014;**54**(3):605-612

[12] Patisson F, Bellot JP, Ablitzer D. Study of moisture transfer during the strand sintering process. Metallurgical Transactions B. 1990;**21**(1):37-47

[13] Aota J, Morin L, Zhuang Q, Clements B. Direct reduced iron production using cold bonded carbon bearing pellets part l—Laboratory metallization. Ironmaking and Steelmaking. 2006;**33**(5):426-428

[14] Zhuang Q, Clements B, Aota J, Morin L. DRI production using cold bonded carbon bearing pellets part 2—Rotary kiln process modelling. Ironmaking and Steelmaking. 2006;**33**(5):429-432

[15] Zhu DQ, Mendes V, Chun TJ, Pan J, Li QH, Li J, et al. Direct reduction behaviors of composite binder magnetite pellets in coal-based grate-rotary kiln process. ISIJ International. 2011;**51**(2):214-219

[16] Wu ZL, Shi LL. Test of coal mixing for direct reduction of iron by one-step process. Mining and Metallurgical Engineering. 2011;**31**(1):63-69 (in Chinese)

[17] Lu ZH, Chen Y, Chen DM, Li SQ. Experimental research on rotary kiln one-step process DRI productions. Journal of Iron and Steel Research International. 2011;**23**(5):11-14 (in Chinese)

[18] Zhu DQ, Qiu GZ, Jiang T. An innovative process for direct reduction of cold-bound pellets from iron concentrate with a coal-based rotary kiln. Journal of Central South University of Technology. 2000; **7**(2):68-71

[19] Sun K, Lu WK. Atheoretical investigation of kinetics and mechanisms of ironore reduction in an ore/coal composite. ISIJ International. 1999;**39**(2):123-129

[20] Nascimento RC, Mourão MB, Capocchi JDH. Kinetics and catastrophic swelling during reduction of iron ore in carbon bearing pellets. Ironmaking and Steelmaking. 1999;**26**(3):182-186

[21] Sharma MK, Solanki V, Roy GG, Sen PK. Study of reduction behaviour of prefabricated iron ore–graphite/ coal composite pellets in rotary hearth furnace. Ironmaking and Steelmaking. 2013;**40**(8):590-597

[22] Gupta RC, Gautam JP. The effect of additives and reductants on the strength of reduced iron ore pellet. ISIJ International. 2003; **43**(12):1913-1918

[23] Tsujihata K, Mitoma I, Fukagawa Y, Hashimoto S, Toda H. Apparatus for operating a shaft furnace by detecting the falling speed of the charge. Transactions of the Iron and Steel Institute of Japan. US3581070[P]. 1971.

[24] Long H, Bai MH, Jia YZ, Liang DL, Ren SB. Investigation of factors affecting drying characteristics of pellets made from iron-bearing converter sludge. Ironmaking and Steelmaking. 2016;**43**(3):1-7. DOI: 10.1080/03019233.2016.1269039

[25] Long H, Jia YZ, Liang DL, et al. Effect of parameters on reduction behaviour of preheated converter sludge pellets in grate-rotary kiln process. Ironmaking & Steelmaking. 2017;**3**:1-6

[26] Long H. Research of reduction and movement behavior of gas-solid in the hydrogen shaft furnace [doctoral dissertation]. Yanshan University; 2018. pp. 88-101

[27] Keran VP, Baker AC, Ridley AJ, et al. The direct reduction corporation's process technology. In: G.N., ISS Ironmaking Conference Proceedings. Vol. 39. 1980. pp. 412-419

[28] Zhu DQ, Li J, Qiu GZ, et al. One-step process for direct reduction of Xinjiang magnetite concentrate. Journal of Central South University. 2007;**03**:421-427

[29] Ma J. The equipments reform and the technique operation principal points of one-step coal based grate-kiln dr process. Sintering and Pelletizing. 2003;**28**(2):42-47

[30] Lepinski JA. The FASTMET direct reduction process. In: ISS Ironmaking Conference Proceedings, Vol. 52. 1993. pp. 349-352

[31] Miyagawa K, Meissner DC. Development of the FASTMET as a new direct reduction process. In: 1998 ISS Ironmaking Conference Proceedings. Vol. 57. 1998. pp. 877-881

[32] Lehmkuler HJ, Hofmann W, Fontana P, De Marchi G. INMETCO process: An attractive solution for

coal based direct reduction of ore fines. In: International Conference on Pre- Reduced Pellets and Europe, Milan, Italy. The International Metals Reclamation Company. September 23-24, 1996

[33] Gao G, Li JX, Long HM, Wang P, Wei RF. Experimental study on improving strength of dust and sludge carbon-containing pellets after drying. Journal of Anhui University of Technology(Natural Science). 2011;**28**:319-324 (in Chinese)

[34] Tsukerman T, Duchesne C, Hodouin D. On the drying rates of individual iron oxide pellets. International Journal of Mineral Processing. 2007;**83**:99-115

[35] Feng JX, Yang SH. Experiment research into hot air cross-flow drying on carbon-containing pellets. Industrial Heating. 2007;**36**(1):27-30 (in Chinese)

[36] Feng JX, Liang KL, Zhang C, Xu JH, Zhang YM, Yang JB. Development and application of thermal mathematical model of iron ore pellet bed in grate. Journal of Shanghai Jiaotong University (Science). 2011;**16**(3):312-315

[37] Umadevi T, Prachethan Kumar P, Kumar P, Lobo NF, Ranjan M. Investigation of factors affecting pellet strength in straight grate induration machine. Ironmaking and Steelmaking. 2008;**35**(5):321-326

Section 4

Comprehensive Utilization of Red Mud

Chapter 6

The Comprehensive Utilisation of Red Mud Utilisation in Blast Furnace

Andrey Dmitriev

Abstract

State-of-the-art formation of red mud during industrial processing of bauxite in the Sverdlovsk region (Russian Federation) is presented. Red mud chemical composition is presented, and an analysis of existing ways in which they are utilised is executed. In the Institute of Metallurgy of the Ural Branch of the Russian Academy of Sciences, red mud is utilised by introducing it into the charge for the production of iron ore sinter and pellets following the use of sinter and pellets in the blast furnace charge. Metallurgical properties of sinter and pellets (reducibility, strength, softening and melting temperatures) with different contents of red mud in iron ore raw materials are also presented, including the technology of red mud usage in ferrous metallurgy carried out through industrial and laboratorial tests. Additionally, the main technical and economic indicators of blast furnace smelting (productivity, coke consumption, chemical composition of pig iron and slag, etc.) are presented. The possibility and expediency of utilisation of red mud in a blast furnace are shown.

Keywords: red mud, agglomerate, pellets, metallurgical characteristics, blast furnace smelting

1. Introduction

Red mud is one of the mass wastes of the aluminium industry. The exit of red mud fluctuates, depending on the structure and properties of processed bauxite raw materials, within 0.8–1.2 kg per tonne of aluminium. In the world, more than 50 million tonnes of red mud are dumped in dumps or reservoirs each year.

In 60 years of operation of the aluminous shop of the Ural Aluminium Plant, three mud storage facilities have been constructed using more than 500 hectares of area with over 63 million tonnes of red mud. A similar picture is seen with the Bogoslovsk Aluminium Plant: mud storage facilities occupy a space of more than 400 hectares with more than 40 million tonnes of mud.

Currently, about 600 million tonnes of red mud are saved in Russia, and this quantity increases annually by 5–8 million tonnes. In the Sverdlovsk region, mud storage facilities managing more than 137 million tonnes are necessary, and 3 million tonnes of dangerous waste are formed every year.

Negative environmental impact has proven the urgency of the problem of red mud usage. Moreover, there are useful components within the contents of red mud at the industrial level that increase prospects of its utilisation. However, the existence and influence of other red mud components that may variously impact technological processes in the blast furnace, structure and properties of products from blast furnace smelting still require detailed study.

2. State of the art

In works [1, 2] the technological expediency of using red mud as a ferriferous additive in blast furnace smelting has been shown. However, excepting the iron generally provided in the form of Fe_2O_3, which contains 15–20% $A1_2O_3$, the ratio of CaO/SiO_2 is close to one, which contains to 4–5% of MgO. Existence of such a quantity and a combination of slag-forming components allows red mud to be used as an additive to correct properties of final slags on physical and chemical properties.

In work [3] it is noted that the iron oxide accounts for about 50 wt% in the red mud; therefore, it is logical to consider the possibilities of red mud usage in burden structure for receiving blast furnace agglomerates. However, it is also necessary to consider that this material considerably concedes to processed iron ores according to the content of iron. Its application undoubtedly reduces productivity of blast furnaces and increases coke consumption. The specified minus is blocked by pluses from essential increases in the durability of the agglomerate. Red mud is found to prevent polymorphism of two-calcium silicate in the structure of the agglomerate. It also leads to the elimination of internal tension destroying the agglomerate. A reduction of the fraction of 0–5 mm per 1% promotes a gain of productivity of the blast furnace per 1% with the same economy of consumption as the expensive blast furnace coke. The binding properties of red mud are also revealed. The burden pelletising and its gas permeability improve. Productivity of sintering machines raises by 5–10% without capital expenditure. If the guantity of small fraction in the agglomerate is reduced by 3–5%, and reduction in the content of iron averages about 0.5%, the summary of technological efficiency in blast furnace production has a gain of 1.2–2.5% on production of cast iron and 1.5–1.8% on economy of blast furnace coke. In addition, the red mud input in the structure burden agglomerate will form a hardened aluminoferrite ligament of agglomerate and pellets. The exit of 0–0.5 mm during the heating and reduction of the agglomerate and pellets in blast furnaces is reduced by 20–40%. Therefore, iron losses with dust taken out from blast furnaces decrease and productivity grows.

In work [4] a negative factor of the ecology of the Russian blast furnace production is noted: the environmental harm from large dust output caused by low mechanical strength and durability in the reduction of blast furnace agglomerates and pellets. This shortcoming can be eliminated through the initial burden of the waste of aluminous production—red mud. Still, it can be hampered by the negative consequences of a reduction in the content of iron in blast furnace burdens. Calculations prove that the negative effect is compensated for by a reduction in the number of small-sized fractions in blast furnace burdens. Domestic agglomerates have been chosen for an assessment of the influence of additive red mud on the increase in productivity and reduction of coke consumption: Bakal with minimum 43.8%, Kachkanar with maximum 60.0% iron content and the Magnitogorsk agglomerate with an average iron content of 53.10%. Calculations establish an introduction in agglomerate burdens of 2% red mud (with mass percentages: 30.0–50.0 Fe_2O_3, 12.0–15.0 Al_2O_3, 4.0–7.0 SiO_2, 5.0–8.0 CaO, 4.0–5.0 TiO_2, 2.0–4.0 Na_2O, 2–3 others, 8–10 losses on calcination), taking into account the reduction of the content of a trifle that allows an increase in the productivity of blast furnaces in the range of 0.67–3.67%, with economy of 0.23–1.73% coke. What's more, inputting red mud in the structure of the blast furnace burden allows additional technological effects at the expense of education protective scar in a high-temperature zone of blast furnace lining with the raised content in blast furnace slag of hard titanium dioxide.

Work [5] notes a thermal agglomeration of ores and concentrates on the removal of harmful impurities of sulphur and phosphorus. It is useful that red mud is used for the improvement of the quality of iron ore agglomerates and pellets. In this work, it is established that in the agglomeration of red mud in aluminous production, it is possible to delete not less than 45% of alkalis and 65% of sulphur from the mud at an optimum consumption of solid fuel; additionally, by the oxidation-reduction roasting of pellets from red mud, it is possible to reduce the content of alkali, sulphur and phosphorus, respectively, by 58–60%, 31–38% and 10–15%. Sintering and roasting are carried out in semi-factory bowls with a 420 mm diameter, while the oxidising roasting of pellets from red mud is carried out at 1000–1200°C and reduction roasting at 1100–1200°C. Coke is used as a reducing agent at a coarseness of <1 mm, which was loaded into a crucible together with pellets in a mass ratio of 1:1 that provided protection of pellets against oxidation. After oxidising and reduction, roasting pellets were exposed to phase and chemical analysis.

Work [6] performs the development of technology for the Ural Aluminium Plant's red mud enrichment by receiving an iron concentrate of FeO = 45–50%. The structure of red mud minerals was defined when studying a polished section under a microscope using the programme "the Mineral-7". The Ural Aluminium Plant red mud enrichment methods are mainly the magnetic and gravitational methods, as well as the physical and chemical methods; a product attrition in the rotor-pulsating device with participation of processes of cavitation from surface processing by hexametaphosphate for a flocculus dispersion has been chosen. It is revealed that the general mineral content of iron in studied tests of the Ural Aluminium Plant makes 53 and 15% from chamosite with an iron content making 18%. In the structure of red mud, there is a significant amount of amorphous formations (flocculus) in which grains of iron-bearing minerals are included. The study of material structure has revealed the following directions for red mud enrichment: attrition; use of dispergators and rotor-pulsating units for flocculus disaggregation of initial red mud; high-gradient magnetic separation for separation of haematite, chamosite and goethite; and gravitational enrichment on a concentration table for operational development of magnetic products.

The results of research on properties of the monomineral fractions allocated from red mud are given in work [7]. The main fractions of red mud are haematite, chamosite and calcite. These fractions have been received by enrichment methods. At monomineral fractions the material structure and physical characteristics (signs), such as a specific magnetic susceptibility and true density, have been defined, and the factor of visibility between separate fractions on each of the physical signs is calculated. The maximum values of factors of visibility 3.75 and 2.3 on a specific magnetic susceptibility have been received, respectively, between haematite and calcite and haematite and chamosite. It has allowed the prediction of the first operations of the technological scheme of enrichment and the main and recleaning high-gradient magnetic separation with an induction of a magnetic field of 1.4–1.5 Tesla, which have been received when studying the dependence of specific magnetic moments of monomineral fractions from the set induction of a field. Further operations of the scheme (gravitational) have been predicted on the basis of the analysis of fraction visibility factors and checked with the use of a concentration table and Knelson separator. The results of research studies issue the optimum technological scheme of red mud enrichment as receiving an iron concentrate with a 50% general iron content and a 15% output.

Work [8] shows that the additive-enriched red mud in quantities of 4% in the high-basicity agglomerate burden of Kachkanarsky Mining and Processing

Integrated Works leads to growth in the quality of the agglomerate and slightly influences indicators of blast furnace smelting; this means that enriching red mud additives in agglomerative production with subsequent blast furnace repartition is a perspective technology for their utilisation. As a positive influence of red mud additives to agglomerates, the increase in the content of MgO in the final slags with 9.1 to 11.2–11.5% is noted as a slight improvement to the negative influence of mud on properties of slags. The increase of magnesia in the final slags, especially when processing high-fluxed agglomerates, leads to increased stability in their phase structure, particularly regarding areas on the border of melilite and spinel slags displaced towards pure spinel, i.e. in areas of more stable temperatures in the beginning of crystallisation. Influence of the specified factor has allowed properties of slag to be saved at the required level—despite an increase in the content of alumina in slag—and has weakened influence of red mud on its properties.

Work [9] studies the structure, thermal characteristics and composition of red mud in the current production of the Ural Aluminium Plant with the help of spectral methods (a mass fraction on solids: 46.7% Fe_2O_3, 12.8% Al_2O_3, 14.5% SiO_2, 10.85% CaO, 4.7% TiO_2, 4.7% Na_2O, 1.125% others). X-ray analysis is carried out on a diffractometer by X'Pert PRO (PANanalitical, the Netherlands) in monochromatic CuKa radiation. For the purpose of specification of the search for iron atoms in different minerals, samples of red mud have been investigated by a method of Mössbauer spectroscopy. In work MS-1104Em, a spectrometer with a source of Co57 in a rhodium matrix was used. The isomerous shift was relatively defined α-Fe. For high-quality determination of the sizes of mineral phases and distribution of elements in them, the samples of red mud have been studied on the electron microscope JXA-8100 (JEOL, Japan) with a power dispersive system, INCA Energy 400. Extant research of red mud structures shows that from the point of view of iron ore materials, the material represents a difficult system of dispersed and ultradispersed crystals of haematite surrounded by ferriferous phases in the form of a crystal phase of chamosite and amorphous iron-silica-alumina phases. Enrichment of such material through traditional means is represented as difficult to realise.

Work [10] establishes the basis of studying red mud structures in the current production of the Ural Aluminium Plant subject to reduction roasting and attempts to define the expediency of magnetic separation of roasting production on the basis of the results of complex research. For this purpose, briquetted samples of red mud in the current production of the Ural Aluminium Plant are subjected to roasting in a weak reduction environment (a gas mix CO_2 (95%) and H_2 (5%)) and the reduction environment (H_2 (100%)) at 810°C. The completed research study on red mud structure after heat treatment in reduction environments and analysis of similar research has shown that achieving effective magnetic separation will only be possible under conditions that develop physical and chemical impacts on the growth of crystals in the magnetic phases, which form during reduction roasting within the range of 800–1000°C.

The existence of alkalis in red mud represents danger for the technological process of agglomeration, especially for blast furnace smelting technology as it has harmful effects on the flameproof lining of the blast furnace.

In 2011, experts from the United Company RUSAL, together with scientists from "Uralpromenergoproyekt", Institute of Solid State Chemistry of the Ural Branch of the Russian Academy of Sciences, Institute of Metallurgy of the Ural Branch of the Russian Academy of Sciences and the Moscow Institute of Steel and Alloys, have carried out a number of research projects on red mud and its properties with

No.	Fe	FeO	CaO	MgO	SiO_2	TiO_2	Al_2O_3	MnO	P_2O_5	SO_3	R_2O	LC*
Mud 1	28.36	5.88	21.32	0.63	9.24	4.14	12.35	0.30	0.60	0.90	0.76	10.30
Mud 2	27.80	5.79	20.9	0.8	9.85	4.06	11.9	0.35	0.31	0.36	—	11.00

**LC—Losses on calcination.*

Table 1.
Chemical composition of red mud (%).

the aim of creating technology for its complex processing, including projects on dealkalisation, dehydration, enrichment and extraction of rare earth metals. Results of research studies are based on the creation of trial experimental industrial plants for red mud processing by calculating the productivity of 200,000 tonnes of mud in a year at the Ural Aluminium Plant.

The chemical composition of red mud in the current production of the Ural Aluminium Plant is given in **Table 1** (Mud 1), and alkaline-free red mud is listed after the above-stated plant (Mud 2).

3. Laboratory studies and industrial tests

Sintering was carried out on the agglomerative cap of a 320 mm diameter, established in the JSC Uralmekhanobr laboratory. The structure of alkaline-free red mud is given in **Table 2**.

The structure of agglomerative burden for the agglomerative factory of JSC Kachkanarsky Mining and Processing Integrated Works is given in **Table 2** (LC—losses on calcination).

After carrying out sintering-ready agglomerate three times, dumped from a height of 3 m according to GOST 25471-82, granulometric composition of the received agglomerate was defined, an output of suitable, drum-type durability and an abradability in accordance with GOST 15137-77. Agglomerate sampling for all types of tests was carried out in accordance with GOST 26136-84. The metallurgical properties of agglomerate, reducibility and a durability indicator at reduction were defined in accordance with GOST 19575-84. Results of tests are given in **Tables 3** and **4**.

For production pellets, Kachkanarsky Mining and Processing Integrated Works concentrates are bentonite and red mud. The chemical composition of materials is shown in **Table 5** (LC—losses on calcination). The content of particles of class in concentrate makes less than 0.071 mm—90 and 100% in bentonite and red mud.

Material	Weight fraction (%)									
	Fe	FeO	SiO_2	CaO	MgO	Al_2O_3	P	S	C	LC
Iron ore concentrate	62.1	28.1	3.6	1.06	2.06	2.44	<0.01	0.025	—	2.33
Limestone	—	—	1.39	53.2	0.32	0.71	—	—	—	42.4
Coke	—	—	—	—	—	—	—	0.54	83.8	83.8
Ash of coke	6.13	1.08	48.0	8.02	1.96	24.4	0.32	1.34	—	—

Table 2.
Chemical composition of components of agglomerative burden.

No.	Indicator/red mud (%)	0	1	2	3
1	Durability (%)	68.73	72.93	67.07	67.73
2	Abrasiveness (%)	7.13	6.07	11.00	8.27
3	Specific capacity (t/(m^2 hour))	1.13	1.07	0.65	0.86
4	Output of suitable (%)	76.83	71.32	64.08	69.68

Table 3.
Results of tests on agglomerate sintering with additives of red mud (%).

Content of red mud (%)	GOST 15137-77		GOST 17212-84	GOST 19575-84	GOST 26517-85
	Durability (+5 mm, %)	Abradability (−0.5 mm, %)	Reducibility (%)	Durability by Linder	Temperature range (°C)
0	68.7	7.1	65.2	36.5	1153–1266
1	72.9	6.1	73.7	27.4	1178–1334
2	67.1	11.0	69.6	31.5	1169–1313
3	67.7	8.27	69.0	32.1	1168–1309

Table 4.
Metallurgical properties of iron ore agglomerate when using red mud.

Material/element	Fe	FeO	Fe_2O_3	SiO_2	CaO	MgO
Kachkanarsky concentrate	61.9	28.3	57.0	3.93	1.10	2.68
Bentonite	3.6	1.3	3.7	58.5	2.43	2.86
Red mud	27.7	5.79	33.14	7.24	19.9	0.74
Material/element	Al_2O_3	TiO_2	S	LC	Σ	
Kachkanarsky concentrate	2.19	2.75	—	1.0	98.95	
Bentonite	18.3	—	—	5.0	92.09	
Red mud	12.8	—	0.36	11.1	91.07	

Table 5.
Chemical composition of burden components for receiving pellets with different contents of red mud (%).

Heat treatment of dry pellets was carried out in the vertical tubular furnace with height of isothermal zone (100–120 mm). Characteristics of the crude and burned pellets are shown in **Tables 6–8** (1—pellets with 0.7% of bentonite, 2—pellets with 0.7% of bentonite and 2% red mud, 3—pellets with 0.7% of bentonite and 4 mass % red mud, and 4—pellets with 0.7 mass % of bentonite and 6 mass % red mud).

Trial tests of production of agglomerate were carried out in the conditions of JSC Kachkanarsky Mining and Processing Integrated Works. Results are shown in **Table 9**.

Trial tests of production of pellets were carried out in the conditions of JSC Kachkanarsky Mining and Processing Integrated Works GOK. Results are shown in **Tables 10–12** (1—0.22% of binding, 1% red mud; 2—0.22% of binder, 2% red mud; 3—binder 0.44%, 4—1% red mud without binder, 5—2% red mud without binder, and 6—0.44% binder).

No. of test	Binding additives		Properties of pellets				
	Bentonite (%)	Red mud (%)	Damp pellets				Pressure (kg/dry pellet)
			Moisture (%)	Pressure (kg/ pellet)	Dropping (3 m, times)	Dropping (1 m, times)	
1	0.7	—	6.87	1.15	4.0	1.0	2.81
2	0.7	2.0	7.0	1.25	4.2	1.0	2.06
3	0.7	4.0	6.85	1.15	4.5	1.0	1.84
4	0.7	6.0	7.0	1.36	4.5	1.0	1.72
5	—	2.0	6.34	1.13	2.9	0	0.57
6	—	6.0	6.9	1.31	3.6	1.0	1.0
7	0.35	6.0	7.0	1.32	3.9	1.0	1.55
8	1.0	6.0	7.0	1.88	7.4	1.0	3.45

Table 6.
Strengthening characteristics of nonfluxed pellets with different contents of binding additives in the form of bentonite and red mud.

No.	Heat treatment conditions				Durability of burned pellets (kg/pellet)	FeO (%)
	Part of layer	Time lag at 1050°C (minutes)	Roast			
			t (°C)	τ (minutes)		
1	Top	5	1300	6	325	4.17
	Middle	3	1270	3	317	2.92
	Bottom	—	1230	1	260	1.64
	Average	—	—	—	301	2.91
2	Top	5	1300	6	334	3.62
	Middle	3	1270	3	365	2.83
	Bottom	—	1230	1	330	1.89
	Average	—	—	—	343	2.78
3	Top	5	1300	6	300	3.85
	Middle	3	1270	3	376	2.2
	Bottom	—	1230	1	308	1.53
	Average	—	—	—	328	2.52
4	Top	5	1300	6	317	3.3
	Middle	3	1270	3	303	1.93
	Bottom	—	1230	1	306	1.22
	Average	—	—	—	309	2.15

Table 7.
Heat treatment conditions on air of nonfluxed pellets with additives of red mud, their strengthening characteristics and chemical composition.

No.	Binding additives		Properties of pellets					
	Bentonite (%)	Red mud (%)	Damp pellets				Pressure (kg/dry pellet)	Pressure (kg/roasting pellet)
			Moisture (%)	Pressure (kg/pellet)	Dropping (3 m, times)	Dropping (1 m, times)		
1	0.7	—	6.70	1.15	4.0	1.0	2.26	301
2	0.7	2.0	6.75	1.15	4.2	1.0	2.06	343
3	0.7	4.0	6.73	1.10	4.5	1.0	1.74	328
4	0.7	6.0	6.70	1.10	4.6	1.0	1.72	309

Table 8.
Metallurgical properties of iron ore pellets when using alkaline-free red mud.

Experience condition/burden	Base	Fe 53.3% (basicity 2.1)	Fe 54.0% (basicity 1.9)	Fe 54.0% (basicity 1.8)
Concentrate (kg)	15.0	15.0	15.00	15.00
Red mud (kg)	—	1.10	1.15	1.25
Red mud (%)	—	7.00	7.70	8.30
Limestone (kg)	2.85	2.86	2.48	2.29
Fuel (kg)	1.00	1.00	1.00	1.00
Backstock (kg)	6.00	6.00	6.00	6.00
Water (l)	2.40	2.40	2.4	2.15
Height of layer (mm)	280	280	280	280
Moisture after pelletisation (%)	7.80	7.6	8.4	7.50
Moisture after heating (%)	7.00	7.1	8.0	7.0
Burden temperature (°C)	80.0	81.0	81.0	81.0
Pelletisation degree (%)	52.8	43.4	59.6	55.8
Mechanical durability				
+5 mm (%)	56.8	49.9	58.8	59.8
-0.5 mm (%)	5.2	9.1	6.6	8.0
Granulometric structure (%): +40 mm	1.0	2.6	3.5	4.2
+25 mm	8.0	4.2	10	12.6
+10 mm	36.1	24.2	33.5	40
+5 mm	26.7	24.9	24.9	15.0
-5 mm	28.2	44.1	28.1	28.2
C (burden) (%)	4.49	4.66	3.89	3.72
Fe (agglomerate) (%)	53.9	53.3	53.9	54.0
FeO (agglomerate) (%)	9.1	8.17	10.68	9.29
Basicity (shares of units)	2.01	2.06	1.98	1.78
Content in agglomerate (%)				
C	0.11	0.22	0.36	0.11
V_2O_5	0.51	0.48	0.47	0.49
MgO	2.94	2.73	2.48	2.68
Al_2O_3	2.38	2.64	3.43	3.10

Experience condition/burden	Base	Fe 53.3% (basicity 2.1)	Fe 54.0% (basicity 1.9)	Fe 54.0% (basicity 1.8)
TiO_2	2.53	2.53	2.65	2.69
S	0.013	0.016	0.023	0.016

Table 9.
Test results of agglomerate production.

No.	Burden materials						
	Concentrate		Binder			Burden	
	Moisture (%)	Dissemination (%)	Moisture (%)	Dissemination (%)	Swelling ability (times)	Moisture (%)	Dissemination (%)
1	9.3	95.4	4.1	93.2	7.6	9.15	95.5
2	9.3	95.4	4.1	93.2	7.6	9.45	95.5
3	9.3	95.4	—	—	—	—	—
4	8.95	94.2	—	—	—	9.0	93.6
5	8.95	94.2	—	—	—	8.95	93.4
6	—	—	—	—	—	—	—

Table 10.
The industrial test results of burden materials.

No.	Quality of damp pellets									
	Moisture (%)	Plasticity (times)	Durability (kg/pellet)		Granulometric structure (%)					
			Damp	Dry	+20	+15	+12	+10	+8	–8
1	8.0	5.1	1.23	3.84	13.5	40.8	21.8	18.5	4.2	1.2
2	8.85	5.8	1.46	4.0	4.2	39.1	46.0	8.8	1.6	0.3
3	8.85	9.1	1.3	5.0	1.5	30.2	47.6	15.9	3.2	1.6
4	9.35	13.8	1.06	1.37	0.0	12.6	19.7	24.9	36.1	6.7
5	9.55	4.3	0.7	1.54	1.9	10.3	20.8	32.0	29.0	6.0
6	8.8	6.8	1.12	3.31	1.3	35.0	45.0	12.5	5.0	1.2

Table 11.
The industrial test results of pellets.

No.	Burned pellets			
	Compressive strength (kg/pellet)	Laboratory mill		FeO, %
		More 5.0 mm	Less 0.5 mm	
1	231.4	96.2	2.5	2.7
2	245.2	97.0	2.7	2.5
3	211.0	98.5	1.4	2.0
4	251.7	96.0	3.0	2.3
5	257.6	97.5	2.5	2.1
6	255.0	97.0	2.0	3.3

Table 12.
The industrial test results of burned pellets.

4. Calculated research

According to the received properties of experimental agglomerates in carrying out laboratory research with an introduction in burden of blast furnace, 1 and 3% of red mud have been calculated. Results of comparative calculations by means of mathematical models [11] are shown in **Table 13** [12].

The analysis of results shows that positive improvement in durability of the agglomerate when using red mud is practically levelled by a decrease in the general content of iron in the burden. To save the content of iron at the level of its base values in the burden of blast furnace smelting, it is possible to use several options: a decrease in basicity of the high-basicity agglomerate, along with a simultaneous raise in the relation of CaO/SiO_2 in Staflux, the addition of a new iron-containing component in the agglomerative burden (scale), and the replacement of part of the bentonite in mud in the production of pellets.

Indices	Unit of measure	Base	Red mud (1%)	Red mud (3%)
Useful volume of furnace	m^3	2200	2200	2200
Productivity	t/day	6241	6235	6141
General consumption of ore	kg/t pig iron	1712.9	1728.8	1742.2
Pellets	kg/t pig iron	907.8	916.3	923.4
Agglomerate, red mud (0%)	kg/t pig iron	650.9	0.0	0.0
Agglomerate, red mud (1%)	kg/t pig iron	0.0	657.0	0.0
Agglomerate, red mud (3%)	kg/t pig iron	0.0	0.0	662.0
Staflux	kg/t pig iron	154.2	155.6	156.8
Coke	kg/t pig iron	396.6	394.5	397.9
Dust exit	kg/t pig iron	39.1	39.4	39.7
Consumption of natural gas	m^3/t pig iron	125.2	125.2	125.2
Blast:				
Temperature	°C	1206	1206	1206
Moisture	g/m^3	8	8	8
Oxygen	%	28.7	28.7	28.7
Consumption	m^3/t pig iron	925.0	924.7	936.7
Top gas:				
Temperature	°C	173.9	176.7	189.5
exit	m^3/t pig iron	1474.5	1469.1	1483.7
CO	%	23.6	23.1	23.1
CO_2	%	21.9	22.4	22.3
H_2	%	9.6	9.5	9.5
N_2	%	44.9	45.1	45.2
Extent of use CO	Shares of units	0.493	0.481	0.480

Indices	Unit of measure	Base	Red mud (1%)	Red mud (3%)
Extent of use H_2	Shares of units	0.464	0.457	0.460
Theoretical temperature of burning	°C	2006	2005	2012
Composition of pig iron:				
Si	%	0.10	0.10	0.10
Ti	%	0.15	0.15	0.5
Mn	%	0.327	0.329	0.346
Cr	%	0.105	0.106	0.112
V	%	0.407	0.415	0.436
S	%	0.025	0.018	0.021
C	%	4.666	4.666	4.666
P	%	0.051	0.052	0.054
Fe	%	94.17	94.16	94.11
Temperature of metal	°C	1450	1450	1450
Slag exit	kg/t pig iron	347.0	359.0	373.0
Composition of slag				
CaO	%	34.20	34.82	33.34
MgO	%	9.34	9.58	9.77
SiO_2	%	27.97	25.27	25.84
Al_2O_3	%	15.83	16.79	17.34
TiO_2	%	10.38	11.32	11.07
MnO	%	0.41	0.41	0.43
Cr_2O_3	%	0.05	0.05	0.06
V_2O_5	%	0.25	0.23	0.26
S	%	0.80	0.77	0.76
R_2O	%	0.57	0.54	0.91
FeO	%	0.61	0.61	0.60
CaO/SiO_2		1.22	1.38	1.29

Table 13.
Calculated indicators of blast furnace smelting of agglomerate with additives of red mud.

5. Conclusion

Laboratory tests, calculations, research, and industrial tests have confirmed the possibility of red mud utilisation as a ferriferous additive (agglomerate) and binding as a bentonite substitute (pellets).

Acknowledgements

This research study was supported by the State Task of Institute of Metallurgy of the Ural Branch of the Russian Academy of Sciences, the Project No. 0396-2015-0081.

Conflict of interest

The author declares no conflict of interest.

Thanks

The author expresses gratitude to Leopold Leontyev, Galina Gazaleeva, Evgeniy Bratygin, Dmitriy Volkov and Yuriy Chesnokov.

Author details

Andrey Dmitriev
Institute of Metallurgy of Ural Branch of Russian Academy of Sciences, Ekaterinburg, Russia

Address all correspondence to: andrey.dmitriev@mail.ru

References

[1] Utkov V. Prospects of development of ways of processing red mud in the USSR and abroad. Moscow: Tsvetmetinformatsiya; 1981. 31 p

[2] Shmorgunenko N, Korneyev V. Complex processing and use of moldboard mud of aluminous production. Metallurgy. 1982. Moscow. 129 p

[3] Trushko V, Utkov V, Bazhin V. Urgency and Possibilities of Complete Processing of Red Mud of Aluminous Productions. Notes of Mining Institute. 2017;**227**:547-553. DOI: 10.25515/PMI.2017.5.547

[4] Trushko V, Utkov V, Sivushov A. Reducing the environmental impact of blast furnaces by means of red mud from alumina production. Steel in Translation. 2017;**48**(8):576-578. DOI: 10.3103/S0967091217080149

[5] Utkov V, Leontyev L, Yakovlev M. Reduction of the content of alkalis, sulfur and phosphorus at thermal okuskovaniye red shlama. Steel. 2013;**2**:12-13

[6] Gazaleeva G, Orlov S, Sopina N, Mushketov A, Anashkin V, Vishnyakov S, Klimentenok G, Petrov S, Kotov O. Influence of material structure of red mud on technological indicators of their enrichment. In: Proceedings of the Fourth International Congress Non-ferrous metals-2012. Krasnoyarsk; 2012. pp. 128-130

[7] Gazaleeva G, Mushketov A, Sopina N, Vlasov I, Uporov S. Selection of the scheme of enrichment of red mud. Non-Ferrous Metals. 7:46-50

[8] Bratygin E, Gazaleeva G, Dmitrieva E, Kalugin Y. Use enriched red mud by production of high-fluxed agglomerate for the purpose of their further processing in blast furnaces. The Bulletin Ferrous Metallurgy. 2013;**1**:30-34

[9] Podgorodetzkij G, Gorbunov V, Korovushkin V, Panov A. Red mud of ural aluminium works flow-line production structure study. Ferrous Metallurgy. 2012;**52**(2):24-29. (In Russ.). DOI: 10.17073/0368-0797-2012-3-24-29

[10] Podgorodetzkij G, Gorbunov V, Korovushkin V, Panov A. Red mud of ural aluminium works flow-line production structure study after thermal treatment in reduction gaseous atmosphere. Ferrous Metallurgy. 2012;**55**(5):8-14. (In Russ.). DOI: 10.17073/0368-0797-2012-5-8-14

[11] Chentsov A, Chesnokov Y, Shavrin S. Balance Logic-Statistical Model of Blast Furnace Process. Yekaterinburg: Russian Academy of Sciences; 2003. 164 p

[12] Yu C, Leontyev L, Sheshukov O, Dmitriev A, Vitkina G, Marshuk L. Pyrometallurgical processing of waste of aluminum production. Bulletin of Magnitogorsk state technical university named after G.I. Nosov. 2013;**3**(43):19-22